Appropriate Technologies for Developing Countries

Prepared by

RICHARD S. ECKAUS
Massachusetts Institute of Technology

for the

PANEL ON APPROPRIATE TECHNOLOGIES FOR
DEVELOPING COUNTRIES
Board on Science and Technology
  for International Development
Commission on International Relations
National Research Council

NATIONAL ACADEMY OF SCIENCES

Washington, D.C.   1977

This report has been prepared by Richard S. Eckaus, Professor of Economics, Massachusetts Institute of Technology, Cambridge, Massachusetts, for a special panel of the Board on Science and Technology for International Development in cooperation with the Office of the Foreign Secretary, National Academy of Engineering, Commission on International Relations, National Research Council, under Contract No. AID/csd-2584, Task Order No. 1, with the Office of Science and Technology, Bureau for Technical Assistance, Agency for International Development, Washington, D.C.

Library of Congress Cataloging in Publication Data

Eckaus, Richard S   1926-
    Appropriate technologies for developing countries.

    "Prepared . . . under contract no. AID/csd-2584, task order no. 1, with the Office of Science and Technology, Bureau for Technical Assistance, Agency for International Development."
    Bibliography: p.
    1. Underdeveloped areas—Technology. 2. Technology transfer. I. National Research Council. Panel on Appropriate Technologies for Developing Countries.
II. Title.
HC59.7.E25   338.91'172'4   76-53284
ISBN 0-309-02602-4

Available from:
Printing and Publishing Office
National Academy of Sciences
2101 Constitution Avenue, N.W.
Washington, D.C. 20418

Printed in the United States of America

PANEL ON APPROPRIATE TECHNOLOGIES FOR DEVELOPING COUNTRIES

Members

BRUCE S. OLD, Vice President, Arthur D. Little, Inc., Cambridge, Massachusetts, *Chairman*

JAMES P. GRANT, Director, Overseas Development Council, Washington, D.C.

CHRISTIAN KRISTOFF, Manager, Corporate Product Planning Group, General Motors Corporation, Detroit, Michigan

FREDERICK T. MOORE, Economic Advisor, Industrial Projects Department, World Bank, Washington, D.C.

FRANCISCO SAGASTI, Science and Technology Policy Instruments (STPI) Project, Lima, Peru

JOSEPH STEPANEK, Consultant, Checchi and Company, Washington, D.C.

SIMÓN TEITEL, Consultant, Office of the Program Advisor, Inter-American Development Bank, Washington, D.C.

Consultant and Report Author

RICHARD S. ECKAUS, Professor, Department of Economics, Massachusetts Institute of Technology, Cambridge, Massachusetts

Staff

JAY J. DAVENPORT, Board on Science and Technology for International Development, Commission on International Relations, National Research Council, *Professional Associate*

HUGH H. MILLER, Office of the Foreign Secretary, National Academy of Engineering, *Executive Secretary*

JULIEN ENGEL, Board on Science and Technology for International Development, Commission on International Relations, National Research Council, *Head, Special Studies*

iii

Preface

Since the 1960's, science and technology have been increasingly recognized as powerful influences in economic growth processes. This is nowhere more evident than in the developing countries, which have in some sectors gained enormously from the use of the technological knowledge accumulated in the industrialized countries. Agricultural and industrial production, the discovery and exploitation of natural resources, and transportation, communications, health care, education, and many other fields have benefited from technical innovations. Yet the transfer of technology to the developing countries has been accompanied by new economic and social problems. While some of these may be inherent in the processes of modernization, others appear to be specifically associated with the characteristics of the technologies transferred from industrialized countries.

By the late 1960's, there was growing concern about apparent incongruities between the goals of the developing countries, their labor supply conditions and other resource endowments, and the technologies these countries were importing. Advanced technologies were characterized by the use of less labor per unit of output, meaning that they not only failed to expand employment as fast as output but, in some instances, may actually have displaced labor and contributed to already substantial unemployment. The new technologies also often seemed to require large plants and equipment and thus appeared to be relatively capital intensive, a strain in countries where capital is unusually scarce. Further, the imported technologies appeared to require the creation of large establishments and to demand new and higher levels of labor skills, to make new social demands on the countries, and to change their economic and social structures. Foreign exchange seemed to become increasingly absorbed in the payment of royalties for licenses for foreign technologies, thereby fostering dependency relationships.

The effects we are describing are said by some development specialists to lack verification; others would differ with this view. There has, of course, been a growing preoccupation within developing countries themselves with the adaptation of technologies to their particular economic and political objectives and conditions. This concern is also manifested by national and international lending and assistance agencies and among many development specialists. Attention has turned to the possibility of finding new technologies more suitable than those now available. However, the difficulties in specifying precisely the characteristics that would make new technologies more suitable are apparent in the variety of names they have been given in anticipation of their discovery. These names include: "labor intensive," "progressive," "capital saving," "village level," and "intermediate." The term "appropriate" has come to be most widely used, perhaps in recognition that there are many conditions determining the degree of suitability of a particular technology to any environment: the political and economic objectives of each country, its social structure and functioning, and the availability and quality of its productive resources.

In the early 1970's, the National Academy of Sciences (NAS)– National Academy of Engineering (NAE) through the Board on Science and Technology for International Development (BOSTID) joined the discussion of scholars and practitioners who were caught up in the task of analyzing criteria for technological choices for development. The U.S. Agency for International Development (AID) requested BOSTID to study the issues inherent in the concept of "appropriate" technologies.

In accordance with its usual practice, BOSTID convened a panel broadly representative of current knowledge and practice on problems of technological choice, utilization, adaptation, and transfer to guide the study and produce a report. This was preceded by meetings of several *ad hoc* groups brought together to chart a course of action. These groups were composed of economists, engineers, scientists, development administrators, and others experienced in the problems of technology transfer, utilization, and adaptation. To limit an inquiry that threatened to become as broad as the development process itself, the panel confined its attention to the industrial sector (with a brief look at agriculture) and to making recommendations bearing on the long-term industrial development of the less-developed countries.

The characterization of technology strategies measured against criteria of appropriateness proved to be more formidable than was initially realized. In the end, BOSTID turned to Richard S. Eckaus, Professor of Economics at Massachusetts Institute of Technology, for the demanding task of writing a report that would analyze the interrelationships between technological choices and economic, social, and political aspects of the development process. This report, which has as its primary aim to help decision makers and others become more aware of the complexities and constraints inherent in technological choice, was written by Professor Eckaus on the basis of his own extended study and several rounds of discussion and review with the panel and its members. The report is not meant to be a scholarly treatise or a comprehensive review of the literature. It is an exposition of the considered views of Professor Eckaus discussing the interaction among the many factors essential for the rational selection of a technology.

The views finally expressed in the document are those of Professor Eckaus, with broad concurrence by members of the panel. It should be noted that one member of the panel (see Appendix) and some experts in the larger interested community do disagree, both on some details and on some principles. We feel the report will make a useful contribution if it stimulates further discussion and thereby helps bring about the recognition that technological choices are always made within a political and social context and that, whatever a society's policies or goals, all criteria for making choices bear a price. When costs and benefits are better understood, more rational technological choices are likely to ensue.

As Chairman of the Panel on Appropriate Technologies, I take pleasure in thanking each of the panel members for his patience and cooperation in this endeavor over many months of effort. Special thanks and recognition are owed Professor Eckaus, not only for the scholarship he brought to the preparation of the panel's report but also for his forbearance and dedication.

On behalf of the panel I also wish to thank those who prepared background materials for the report. Although they are not reproduced here, the informal case studies and the panel's discussions with their authors were of great help in clarifying the issues. Contributors to this part of the work included José Giral, of the National Autonomous University of Mexico, on appropriate chemical technologies; W. Paul Strassmann and John S. McConnaughey, of Michigan State University, on housing and residential construction; Gerard K. Boon, of the

College of Mexico, on economic technological behavior in the metal-working industries; Charles Kusik, of Arthur D. Little, Inc., on iron and steel making; Vernon W. Ruttan and Hans Binswanger, of the Agricultural Development Council, on technology transfer and research in agriculture; and Simón Teitel, Mauricio Thomae, Hugh Schwartz, and José Villavicencio, of the Inter-American Development Bank, on acquisition, adaptation, and technology-development experiences of entrepreneurs and decision makers in developing countries.

We are indebted to Gustav Ranis, Economic Growth Center, Yale University, who joined us in our discussions and commented in detail on an early draft of the report. The panel also extends thanks for thoughtful written comments contributed by Nazli Choucri, Department of Political Science, Massachusetts Institute of Technology; Walter Falcon, Director of the Food Research Institute, Stanford University; and Larry E. Westphal, Development Economics Department, World Bank.

Finally, I wish to thank Mercedes Rodriguez-Dickinson, who patiently typed the manuscript for printing; Elizabeth McClure, our helpful editor; and Karen Hladek and Diosdada De Leva for bibliographical assistance.

BRUCE S. OLD
Chairman

Contents

xi

Overview

Development Processes

The burgeoning interest in finding and implementing "appropriate technologies" reflects a recognition of the essential role of technology in development. Technological decisions and the pace of technical change affect all development processes—economic, political, and social—and, in turn, are affected by those processes. However, there is great scope for variation in the relations among technological decisions and development processes and little basis for a belief in technological determinism.

Economic growth is generated from several different sources and in different patterns. In general, the following economic changes are combined in successful development:

- Increases in the amount of resources available;
- Improvements in economic efficiency with which resources are used;
- Technological change that expands the potential productivity of resources;
- Changes in the relative weights in the economy of agriculture, industry, and other sectors; and
- Changes in organizational methods, with a declining role for informal and family enterprise and a widening scope for specialization in production and exchange.

1

Of the many problems that emerge in the course of development, the growth of open unemployment is among the more intractable and socially disruptive. It has its sources in rapid population growth and the displacement of labor. The fundamental economic means for resolving the problems are the acceleration and improvement in growth processes, including the use of technologies now available for the intensive and efficient use of labor. There is little evidence to suggest that major research efforts to find efficient "intermediate" technologies for small-scale village-level production would either be markedly successful or contribute substantially to development.

Criteria of "Appropriateness" of Technology

The use of any particular technology is not an end in itself. The criterion for an "appropriate" technological choice must be found in the essential goals and processes of development. A number of different criteria have been proposed, either implicitly or explicitly. These include the maximization of output, maximization of the availability of consumption goods, maximization of the rate of economic growth, reduction in unemployment, regional development, reduction in balance of payments deficits, greater equity in the distribution of income, promotion of political development (including national self-reliance), and improvement in the quality of life. This last criterion has recently been interpreted as development by means of relatively self-sufficient village or rural activities.

These criteria are not only alternative in the sense of attracting different degrees of support but, in many circumstances, can also be competing. In addition, understanding of the relationships between them and technological decisions is incomplete, especially for the last three criteria.

Sources of Technological Information and Decision Processes

There are a number of different sources of technological information and means of dissemination. These include new research and development, transfer of existing knowledge through technical literature, specialized education, technical consulting and equipment sales, and, finally, adaptation of known methods. The available information does not demonstrate that industrial research and development organizations in developing countries have been markedly successful compared with agricultural research institutions. However, there is insufficient data to judge the relative significance, benefits, and costs of the alternative methods of generating and transferring technological knowledge.

Technological decisions are made under different circumstances and with different motivations that affect the degree to which they achieve any of the criteria of appropriateness. Private enterprises may choose their technologies to maximize profits, but this choice will be consistent with the maximization of total national output and income only if there are perfect markets for resources and products. In actuality, markets have many imperfections, including some degree of market control through monopoly power, the segmentation of markets, and government interference through regulations, taxes, and subsidies. In developing countries, informally organized family enterprise continues to be important and, in general, it does not try to maximize profits in a conventional sense. Government and public enterprise can be directed to pursue any of the criteria of appropriateness, and even private enterprise can be controlled for such purposes. Yet, unless technologies that maximize profits or minimize costs are chosen, continuing subsidization from the government budget is likely to be required. The conflicting goals often posed for government and public enterprises obstruct their achievement of any particular objective.

Technological Choices in Agriculture, Service, and Small-Scale Enterprise

Agriculture and the choice of agricultural technology have a particularly critical role in development given the size of this sector in most developing countries. Much of the agriculture in these countries is organized in family farms whose goals are not the conventional profit-maximizing ones and which, because of their small size, cannot engage in technological experimentation. Nonetheless, peasant farmers have been willing to adopt agricultural innovations when they are demonstrated to be profitable. There have been important innovations in agriculture in high-yielding seed varieties, intensive use of fertilizer, and mechanization. But conflicting reports of the effects of these innovations on employment and income distribution make it impossible to conclude whether, on balance, they displace labor, although in some circumstances that may well happen.

Small-scale and service enterprise appear to be relatively successful in absorbing labor. The technologies used in the latter sector seem to have greater potential for efficient use of labor-intensive techniques.

This may also be true of most small-scale enterprise in general; however, it is difficult to determine whether the labor-intensive methods used are actually efficient.

Conclusions and Policy Recommendations

Unfortunately, little detailed, quantifiable information about the characteristics of technology is readily available in a form necessary for policy formation. Qualitative and verbal descriptions are inadequate. This lack of information means that there is little reason to believe that simple and sweeping proposals for the adoption of particular types of technology will be "appropriate" by important development criteria mentioned earlier. Moreover, no panaceas or quick remedies for the problems of development are to be found in the choice of particular types of "intermediate" technologies.

It is clear from existing studies that there are efficient technological possibilities for using more labor-intensive methods in a number of sectors in the developing countries. These are typically not fully exploited, in part because of distortions in the prices of labor, capital, and other resources sometimes created by misguided government policies to stimulate development. Thus, the first recommendation is to encourage the formation of resource and product price policies that will encourage the use of efficient labor-intensive methods.

While only modest expectations and resource allocations are now warranted, engineering research to attempt to extend the range of efficient technological choices would be worthwhile to help the developing countries benefit from their particular resources. The second recommendation, therefore, is for technological and economic research that would start by first identifying the particular products and/or processes for which such research is likely to be successful.

The third recommendation is to investigate the conditions that appear to have stimulated the adoption of efficient labor-intensive technologies and the methods for effective dissemination of information about such methods. In particular, the role of industrial and trade associations, which in some countries have enabled small-scale enterprise to compete successfully using relatively labor-intensive techniques, should be studied.

In negotiations for the transfer of technology through licensing agreements, buyers and sellers have unequal knowledge and bargaining power. Since these negotiations may, therefore, result in distorted choices and charges, the means for improving the outcome of such licensing negotiations deserve particular attention.

1 Summary

Introduction

Technological decisions and the pace of technical change affect all development processes and, in turn, are affected by them. The combinations and proportions in which labor, material resources, and capital are used influence not only the type and quantity of goods and services produced, but also the distribution of their benefits and the prospects for overall growth. The significance of technological choices made in the course of development extends beyond economics to social structure and political processes as well. The new products and new methods, which are the central features of the technological transformations in the developing countries, represent especially sharp breaks with the past and thus have especially profound consequences. Because they interact extensively and intricately with development processes, technological decisions can also be used as conscious instruments of development policy.

The growing interest in finding and implementing "appropriate" technologies reflects an implicit, if not always explicit, recognition of the essential role of technology in development. This report examines the role of technology in developing countries to determine the content and methods of appropriate technological decisions. Decisions about technology are always specific; they are choices of particular products and production methods. The objective of this report, however, is to arrive at generalizations about the character and consequences of such microeconomic decisions.

5

Development is a complex process that varies among countries, reflecting differences in their social structures and goals, population patterns, natural resource endowments, and capital accumulation. In all cases, development means the alleviation of the terrible poverty that afflicts most of the world, but it requires additional resources as well as increased productivity and a wider variety of goods and services. By necessity, then, development requires new technologies. Development also involves social modernization and political transformations, which both require, and are affected by, new technologies.

Recognition of technology's essential role in development does not imply a technological determinism. Not only can alternative products and methods be chosen, but the wider effects of these choices depend strongly on the political and economic environments in which they are implemented. In particular, technological change is not necessarily beneficial for all development goals. Depending on the circumstances of their introduction and use, technologies that increase resource productivity may, for example, also increase income inequality or social stratification or urban crowding. Although there are many intuitively plausible relationships among technological decisions and economic and social development, formulating policies for technology requires a deeper understanding, than intuition alone provides, and the more rigorous definition of relations.

"The poor have always been with us," but only since World War II has there been general recognition and international concern for the massive problems associated with the poverty of the developing nations. This new attention to development problems has been closely related to the achievement of independence by colonial areas and the emergence of governments that have made economic progress a primary objective. The 1950's and the early 1960's appear, in retrospect, to have been a period of euphoria with respect to development. There was general confidence that the problems of poverty could be overcome and, in particular, that rapid development would follow independence and the implementation of conscious national development policies. Those national policies had their counterparts in the economic assistance programs of the advanced countries, which, by their short-term nature, also implicitly predicted such rapid progress that the need for assistance would be relatively brief.

Progress was indeed rapid in many of the developing countries during the first 15 or 20 years after the end of World War II. In many of these countries, great new construction projects were undertaken; new industries were introduced; and, undoubtedly, growth rates of aggregate and per capita income were far higher than they had previously been, as well as high relative to those of many advanced countries. There was also important progress on other social fronts: educational and medical systems were expanded in many areas, new types of social organizations were introduced, and political participation appeared to increase in many countries.[1]

By the mid-sixties and even earlier in some cases, however, a growing pessimism—perhaps realism—with respect to development prospects emerged in the industrialized countries and in the developing countries, whose growth objectives became increasingly modest. In addition, the governments in some of the developing countries became spectacularly unstable. Though such political instability may not have been characteristic of developing countries alone, it was widely believed in the West that the instability reflected a failure of development efforts. Both the industrialized and the developing countries began to doubt the major themes that had guided investment and foreign trade. A succession of crop failures due to drought in parts of Asia and Africa came to be regarded as evidence, in part, of a mistaken preference in investment policy for industry over agriculture. Several unforeseen and important side effects of major construction projects substantially diminished the hoped-for benefits. The optimism that accompanied the Green Revolution decreased when the expectations for a solution to the world's food problems turned out to be unwarranted. Inflation intensified in some of the countries in which it was endemic and appeared in other areas that had been previously immune. Other development problems persisted, and progress appeared limited on nearly all fronts. This pessimistic view of progress in the developing countries has come to be widely held in the seventies, with only a few countries somewhat controversially identified as "success stories."[2] Partly because of that disillusionment, the economic assistance programs of most industrialized nations, including the United States, have declined in real if not money terms and have received less and less political support. Table 1 tabulates real per capita growth rates in the industrialized and less-developed countries since the 1960's. As those numbers indicate, the permission has not been uniformly justifiable.

In the early, optimistic period of development analysis and policy as well as in the more recent pessimistic days, technology has been attributed a critical role both in resolving and creating development problems. The potential importance of technological decisions for employment and unemployment and the possible conflicts between employment, output, and other social goals of development were recognized, and strong views were formulated early in the economic analysis of development problems. The conflicting strategies of

TABLE 1 Real Rate of Growth of Gross Domestic Product Per Capita (Percent)

Country	1961–65	1966–70	1971–74[a] (Average)
Developing	2.9	3.3	3.8
Africa	2.3	2.3	2.4
Southern Europe	5.6	4.8	5.1
East Asia	2.9	4.7	4.9
Middle East	5.1	4.5	8.3
South Asia	1.3	2.0	−0.6
Western Hemisphere	2.4	2.9	4.1
Industrialized	3.9	3.6	2.8

[a]Preliminary data.
SOURCE: World Bank Annual Report 1975.

Mahatma Gandhi and his political disciple, Jawaharlal Nehru, exemplify two of the ideologies regarding relations between technology and development. Gandhi advocated reliance on small-scale agriculture and village industry and the technologies associated with them. Nehru, on the other hand, was responsible for the first three Indian Five-Year Plans that vastly expanded the heavy-industry sector using modern technologies with all their accoutrements. In the first debates over Indian development policy, it was recognized that the sectoral decisions also implied technical choices with far-reaching consequences for employment.[3]

Most recently, there has been a resurgence of concern with the policy implementations of technological decisions under the headline of discovering and implementing "appropriate technologies" for economic development. Avoidance of widespread unemployment has moved into the forefront of economic issues in many of the developing countries. This problem is related to the rapid population growth rates that accelerated 20 to 30 years ago and that are now beginning to deliver massive numbers of new entrants to the labor forces of many developing countries. If the patterns of labor absorption of the recent past continue into the future, many of the new potential workers will not find jobs. That, in turn, would have important consequences for the distribution of income and wealth unless consciously offset by government tax and expenditure programs.

Such diverse problems as balance of payments difficulties and urbanization can also be related in part to technological decisions. These decisions affect the relative costs and the quantities of imports and exports and, therefore, the characteristics of each country's balance of payments. In addition, patent fees and royalties for imported technologies are a direct expenditure for foreign countries. Similarly, the exploding problems of urbanization in developing countries are associated with the technological choices that affect relative sectoral development. Technological decisions in part also determine the relative expansion of employment opportunities in rural and urban areas, which, in turn, affects the patterns and intensities of internal migration.

Development Processes and Technological Decisions

The current output of goods and services depends on the efficiencies with which labor, capital, and other available resources are used. These are determined in part by the technological choices made from the available alternatives. Technological choices, in turn, depend on the character of economic motivations; on immediate influences such as market prices of outputs and inputs and government taxes and subsidies; and on political influences constraining engineering and economic decisions.

The increases in output necessary to alleviate poverty require more and more resources. Technological change in the production of old products as well as the introduction of new products will also account for a substantial part of the economic growth. The potential contribution of technical change is related to investment rates, however, since new capital formation in many cases is the "carrier" of technological improvements. Investment must be based on either domestic private or public saving or foreign saving, but the substitutability of domestic and foreign saving will depend, in part, on the nature of the technologies employed. Although open and disguised unemployment are increasing problems in many countries, population growth in many other countries with rich resources and low population density will contribute to overall growth. Whether labor-force growth intensifies the unemployment problems in developing countries depends in part on the employment characteristics of the technologies employed. The education and skill requirements of the labor force will also be determined by the technologies adopted.

Although there is controversy over the most desirable patterns, changes in the sectoral composition of production are essential for development. Both agriculture and industry must grow, and, if development is successful, the latter sectors will grow faster at some point to match the changes in patterns of demand that accompany income growth. The precise patterns of sectoral growth and the effects on general development goals will again depend partly on the technologies utilized.

Changes in economic organization will both cause and affect technological decisions as the role of household and small-scale enterprise in the economy changes and as public and private corporate activity expands. Technological change is one of the major agents of social modernization affecting society intimately at all levels. Directly and indirectly, technological decisions will affect the processes of political development and be affected by them. It is to be expected that political considerations will often dominate narrowly defined economic goals in the choices of projects, products, and technical methods.

Criteria of Technological Appropriateness

Since the use of any particular technology is not an end in itself, the criteria of appropriateness for the choice of technology must be found in the goals of development. These goals are concerned not only with the volumes of output and income generated by an economy but also with the way they are produced and distributed among the population; they include, as well, particular patterns of national political change and national independence.

The criterion for choice of technology most commonly applied, either implicitly or explicitly, is net output maximization or cost minimization. Application of this criterion will, in general, achieve both physical and economic efficiency in the use of resources and, thus, will do as well as possible in increasing the size of the total output pie. Yet, this criterion would be satisfied by the operation of private markets only if such markets function perfectly, but there are, in fact, many sources and types of "imperfections."

An alternative criterion, maximizing the availability of consumption goods in the present and the future, may have the same implications as maximizing growth, depending on the index of growth used.

Both of the above criteria may be different from maximizing employment. The latter goal, though often thought of in terms of the number of jobs provided, is, however, an ambiguous expression of social policy. Employment serves a number of functions: output generation, income distribution, social recognition and personal satisfaction, and "taxing" of the potential leisure of the community. Thus, if employment is to be an operational criterion of technical choice, its content must be specified in detail.

While income distribution is an issue of increasing social concern, the relations between technical decisions and the achievement of this particular goal are seldom clear. Regional development is a specific aspect of the income-distribution goal. Again, technologies seldom have specific regional implications, although decisions about regional distribution will, in turn, emphasize particular sectors and particular types of technologies.

Similar observations apply to balance of payments relief, which is often set as still another alternative criterion of technological choice. It has been argued that export-oriented production will contribute more to development than import substitution. The claim is based partly on the judgments that international competition faced by export production will promote efficiency and that there is the potential of achieving economies of large-scale production by selling to world markets. However, governments appear as able to distort export patterns via subsidies and other measures as they are to create and preserve distortions in import-substituting industries.

Political development has also been proposed as a criterion of technological choice. The creation of a national political system can be both an instrument and a result of an interrelated national economy in which large-scale projects undertaken with government initiative and support have a major role. By comparison, self-sufficient village-level production is more consistent with dispersed political power and decision making and with lower capability of the political system to evolve and respond to system demands. However, understanding of the political sources and consequences of technical decisions is still at an early stage.

Improvement of the quality of life has recently been emphasized, in general terms, as a major consideration for technological choice. Proponents of this criterion prefer qualities associated with village or rural life, small-scale activity, self-sufficiency, minimum ecological effects, and equality in income distribution. The desirability of these particular qualities and the extent to which other objectives should be sacrificed to achieve them is clearly a matter of preference. Moreover, little attention has as yet been paid to the trade-offs among these quality of life goals themselves and other development objectives.

The criteria proposed for technological decisions are often competitive as well as complementary. It cannot be presumed that pursuing any one will automatically satisfy all or any of the others. Therefore, before an "appropriate" technology can be chosen, the criterion for choice must be determined and, implicitly or explicitly, other criteria must be rejected. Moreover, choices once made are not automatically feasible. If a criterion other than cost minimization is used for private or for public enterprise that must meet a market test, then the enterprise is not likely to be viable without continued government intervention on its behalf. Thus, widespread use of other criteria will, in turn, require a large-scale government program of taxes and subsidies or regulation for the decisions to be effective.

Technological Alternatives and Information Transfer

Information about the range of available technological alternatives, their precise characteristics, and their implications for the criteria of appropriateness is essential for policy making. Yet such information is not readily available, and it is costly and difficult to acquire. Overall investigations based on existing statistics yield only equivocal results, and case studies are seldom generalizable.

Although the weight of evidence suggests that most producing sectors are not restricted to a single efficient combination of resources, estimates of the range of choice available do not clearly demonstrate that wide and easy substitution is possible. Case studies, mainly in light manufacturing, do indicate that alternative techniques are being used, particularly in activities peripheral to central production processing, but these case studies often do not assure that the alternatives are all efficient. Even less information is available to show how various technologies satisfy distributional or other criteria of appropriateness.

Many sources of technological information exist and are used. These include some relatively inexpensive sources such as technical literature, formal and informal education and training of individuals, as well as licensing of patented production processes and sales of expertise and equipment. The intrinsic measurement problems and limited research make it impossible to evaluate the relative importance and potential substitutability of these sources. Some information on the cost of using the various sources exists, but is so limited that generalizations are not warranted. Adaptations of technology in the course of production may be one of the most common and fruitful sources of technical change, but again little hard information is available.

Determinants of Technological Decisions

Technological choices in developing countries are made by various types of decision makers and under a variety of sources of influence. They will not all necessarily have the same objectives or conform to the same ideas of what is "appropriate." The owners and managers of both national private enterprise and international enterprise will be concerned with their own profits, but the latter will have different opportunities and constraints than the former. Both will be reacting to the incentives of the resource and product markets in which they buy and sell. Their technical decisions will be affected by the influences they exert in their markets and the extent to which such markets are unified and are affected by government tax, subsidy, and regulation policies. It may well be optimal for a developing country as well as for a multinational firm to import technologies from more developed countries when both research and development as well as capital and operating costs are taken into account.

Government corporations may or may not behave like private enterprises, depending on their organization and on the objectives and controls imposed on them. In some cases, government enterprise has been used to pursue particular employment and income goals, although the pursuit has sometimes been piecemeal and inconsistent because clear incentives were lacking.

The official national and international economic assistance agencies, such as the U.S. Agency for International Development and the International Bank for Reconstruction and Development, may have a different conception of the development process than the developing countries themselves. They have used their loan and grant programs to advance strategies and impose conditions of performance that the developing countries would not otherwise have adopted.

The methods of implementing decisions also vary among decision makers. In private enterprises, decisions on technology may be made directly but sometimes require approval of a government licensing agency. Governments can resort to direct regulation and their fiscal system to exert their influence. The official international sources of funds rely on their expertise as well as their power to grant or withhold funds to persuade and guide.

While technological decisions in a developing economy should be coherent and compatible with a particular set of goals, such coherence and compatibility is never fully achieved, just as it is not achieved in the developed countries. Rather, economic policy decisions in the developing economies reflect sets of influences that, to some extent, act at cross-purposes. A major task of development policy is to reduce the frustration and inefficiency associated with inconsistent goals and methods, but this task is never fully accomplished. It is especially difficult to achieve when the choice concerns appropriate technologies, because the knowledge, interests, and operating methods of the different decision makers and sources of influence are often at variance.

Technological Decisions in Agriculture

The special features of agriculture in developing countries derive in part from the diversity of their production conditions. Climate and soil conditions can vary widely, even within relatively small countries, and the farm products and the inputs they require also vary. In addition, there is great diversity in the organization of farm enterprise. The family enterprise, as distinct from capitalist organization, is of particular importance in agriculture in the developing countries. In capitalist enterprise, labor and other resources are purchased to maximize profits, while in family farms, resources and technologies are chosen to maximize the net returns to the family's capital and labor as a whole. A number of different institutions prevail in both the formal and implicit contracts that regulate tenancy, labor obligations, and use of other resources as well as in decision-making processes. These are not all equivalent to the practices of conventional business enterprises and require individual analysis.

The search for technological improvements in agriculture is typically centralized in research stations and is usually publicly sponsored because the scale and time horizon of the research is often far beyond what even large landowners can afford. This centralization of technological research requires a system for disseminating results. That system is formalized in the agricultural extension services, which are also feedback mechanisms that spur research to meet farm problems.

Improved seed varieties, fertilizers, pesticides, mechanical equipment, and improved farming practices have resulted in major "revolutions" in agriculture in the developing countries. When individual farmers adopt these innovations, their own interests dictate the criterion applied, just as is the case of individual enterprise in other sectors. The criterion is usually output maximization or cost minimization. However, government policy can exert powerful influences on decisions of individual farmers through taxes and subsidies as well as through direct controls and the provision of new technologies. Though these policies can be used consciously to direct agricultural development, their effects are often unintended by-products as farmers pursue their own interests.

Technical Decisions in Small Enterprise and the Service Sector

There is impressive evidence that the potential for choice of technologies to permit more intensive use of labor is greater within small-scale enterprise than for larger enterprise in the same sector. However, existing studies have not yet established that this greater employment intensity is consistent with the other criteria of technological appropriateness, in particular the criterion of economic efficiency or cost minimization. Small enterprise operates in an environment with many noncompetitive elements although their significance is difficult to evaluate. Moreover, because it typically uses a relatively high proportion of family labor and other resources not provided through markets, small enterprise does not follow the conventional rules of profit maximization. Even the continuing existence of small firms cannot be taken as conclusive evidence of their ability to meet the test of efficiency and survival in a competitive environment.

Some types of traditional small-scale family enterprise are vertically integrated and completely transform raw materials into final products. Other modern small enterprises have specialized to lower costs and coordinate with larger enterprise. The successful achievement of this goal has been facilitated in some cases by government-assisted trade organizations.

While there are questions about the economic efficiency of small-scale enterprise, such enterprise may be successful when judged by the other standards of appropriateness. Not all "inefficient" types of economic activity are socially acceptable, but small-scale family enterprise is typically well regarded and is frequently given special protection. On the other hand, family enterprise has also been the locus of what are now considered social abuses, for example, the intensive use of children and women in production, or work in unhealthy and dangerous conditions, often resulting from the use of housing and other casual structures for production.

The service sectors cover a particularly varied group of activities. Some need highly trained professionals; other sectors depend mainly on persons with little or no training. In the construction sector, which has a particularly critical role in the development process, a number of alternative technically feasible technologies can often be used. Research on highway construction techniques indicates that the most labor-intensive methods and intermediate technologies are not as economically efficient as the capital-intensive methods, but it may still be possible to develop alternative highway designs for which labor-intensive methods do meet the efficiency criterion. Alternative construction technologies may also have different implications for the satisfaction of the criteria of appropriateness other than cost minimization.

Within the service sector, the health care delivery and educational systems appear to have important technological alternatives. These alternatives also provide different "qualities" of services, and these quality differences make it difficult to judge the acceptability of the technological alternatives by the various criteria of appropriateness.

This is also true of the marketing and storage sectors, which are considered important in the achievement of greater benefits from recent increases in agricultural productivity. The potential contributions of alternative technologies in these sectors remain to be carefully investigated.

Policies for Improving Technical Decisions

Because the amount of information and expertise on which to base policy is limited, a major goal at this point must be to improve the knowledge base.

While many influences affect technological decisions, it is not clear how they operate and what their relative weights are. Thus, no single emphasis or policy can be recommended that will ensure that more appropriate technologies can be discovered for developing countries. A single-minded approach is likely to lead to decisions that are inappropriate by some important development criteria. To be appropriate, technological decisions must be tailored to the individual country, the particular sector, and the particular criterion being pursued. Village-level or intermediate technologies are seldom defined precisely and cannot be regarded as solutions for the problems of development. There is little reason to believe that the intensive search for such "intermediate" technologies will facilitate development and that a substantial diversion of resources to this search is warranted.

Three types of policies can be distinguished: (1) policies that improve the incentives operating in the choice of technology; (2) those that will expand knowledge of technological alternatives particularly suited to developing countries; and (3) institutional changes to improve and lower the cost of disseminating technical information.

To help make the choices of technology more consistent with development goals, the economic and political incentives operating on policy decisions should be examined carefully. Market prices may not in themselves ensure the adoption of appropriate technologies, but they certainly should not favor inappropriate technologies. Projects should be evaluated with correct shadow prices as well. Economic research should be sponsored to improve understanding of the technological choice process; this research should include the study of the costs of information and its relation to the exercise of influence over technological decisions and the study of the efficiency of small-scale enterprise.

To expand knowledge of technological alternatives, task forces of engineers and economists should be formed to generate priority lists of production methods and problems. These lists should indicate the sectors in which it was both of particular importance to extend the range of efficient techniques (especially at relatively low levels of output) and in which research would have reasonable opportunities for success. Research to actually explore new technologies should then be supported on the basis of the priorities established.

The general presumption should be that technological investigations should be done in close conjunction with the potential users of the technologies. Technological information should also be accumulated and disseminated with careful attention to the evaluation of its quality and its economic as well as technical implications. Research should also be sponsored to improve the understanding and effectiveness of the public and private mechanisms customarily used to collect and disseminate technological information.

Although institutional conditions are important in the determination of technological choices, these factors are not well understood and research in this area needs to be expanded. For example, the role of trade associations and government standards in expanding and modernizing small-scale enterprise should be investigated. Since technological decisions and political factors are interdependent in development, These interdependencies deserve more study as well.

At present there is some degree of monopoly power in the differential possession of technological information and project-by-project bargaining. Thus, existing licensing methods can be expected to lead to inefficient uses of technological information. Consideration should be given to finding means for improving the terms on which technological licensing is done.

The body of this report begins in Chapter 2 with a brief survey of development processes, with particular attention to technical influences. This survey is intended only to provide a background for the later more detailed analysis of the relations between technological decisions and economic and other social goals and processes. Alternative criteria for technological appropriateness are described and analyzed in Chapter 3. Chapters 4 and 5 review the factors that determine the general effectiveness of a developing country in making appropriate choices of technology. These factors include both the general, systematic influences that condition choice processes, the decision methods implicit in the functioning of the economy, and the conscious and purposeful use of techniques for choosing among development projects. In Chapter 6, some of the unique features of the development processes in the agricultural sectors are surveyed, and their significance for the choice of appropriate technologies is examined. Chapter 7 considers briefly development processes and technical choice in small-scale enterprise and the service sectors. Finally, Chapter 8 suggests technological, economic, and institutional policies to help improve the coherence of development goals and performance.

REFERENCES AND NOTES

1. There is an enormous literature on development history and policy. A useful survey is provided in A. Maddison (1970) and in H. Chenery et al. (1974). The annual reports of the Organization for Economic Cooperation and Development, Development Assistance Committee and the International Bank for Reconstruction and Development (World Bank) provide up-to-date information.
2. For a skeptical view of the achievements of the post-World War II development efforts in improving the conditions of the poorest people in most of the developing countries, see I. Adelman and C. Morris (1973). The observed patterns of development are also surveyed in H. Chenery and M. Syrquin (1975).
3. J. Lewis (1962).

2 Interaction of Technological Decisions and Development Processes

Introduction

A survey of development processes and the role of technology is undertaken in this chapter to provide a general background for the later, more detailed analysis of technological decisions and related policies.[1] Understanding the relationships in development, however, involves special difficulties. First, there is great diversity among the developing countries. Second, despite recent intensive studies, only a limited amount of relevant information is yet available. Finally, and fundamentally, there is only a limited theoretical understanding of the processes of economic growth and social change involved in development.

Most of the characteristics of the developing countries vary enormously. The geography of the developing areas includes, for example, the Ganges basin and the teeming cities of the Asian subcontinent, the high Andean slopes and valleys, the semidesert of sub-Saharan Africa, the richly endowed and relatively sparsely populated African countries, and the great river valleys of the Mideast. There is also wide diversity in the cultural and political patterns of the societies and in their political systems. There are areas in which tribal organization is still important, areas with a long history of political independence, and countries that have emerged from colonialism only recently. There are economies in which there is a high degree of public intervention and direction, economies that are primarily organized by private enter-

19

prise, and many different mixtures of public and private enterprise and control. This diversity must be fully respected in any analysis and is a major barrier to generalization.

While it is easy to recognize the heterogeneity of the developing countries, it is difficult to document in detail the characteristics of any one of them. Their statistical services are often relatively new; scholarly investigations are burgeoning in some places but just beginning in others, and they still do not add up to a comprehensive view of any of the countries.

It is especially difficult to obtain those quantitative measurements of the characteristics of technologies that are necessary to determine the economic and social implications of specific methods. While the ultimate goal is to be able to generalize about the role of technology in the development process, these generalizations must be based on the accumulated knowledge of particular technologies. For this purpose, qualitative and verbal descriptions are inadequate, although frequently they are the only basis presented for policy propositions. Quantitative technical information is lacking for several reasons. Technical decisions are made at the plant and process levels, and the collection of information at that level is a time-consuming and costly process. In addition, technical data in the quantitative form desired are frequently at the heart of the competitive advantage of particular plants and are, therefore, regarded as privileged information.

Although theoretical understanding of development processes has undoubtedly progressed, it is necessary to be modest about what has been achieved. Development is a flourishing field of specialization in political science and economics, but advances are achieved through slow accretion rather than spectacular breakthroughs.

Of all the influences on development, the economic influences of technological decisions can be most thoroughly traced, and they will be emphasized in this survey. The character of other political and social transformations in the development process and their relation to technological decisions will then be discussed briefly.

Efficiency and Technological Change as Sources of Growth

One of the most obvious features of the economic life in the less-developed areas is the strikingly "old-fashioned" technology often employed. While "modern" facilities are being rapidly created and are expanding beyond industrial enclaves in some of the more successful cases, traditional techniques still prevail in many sectors. No special census or accounting of inputs is needed to see that a bullock cart is different from a truck and that handsaws are different from power tools. Neither is special insight needed to understand that a person driving a truck or using a power saw can accomplish more than someone guiding a bullock cart or using a handsaw. And there is little doubt that higher productivity of individual workers is one of the essential requisites of development.

Yet it does not follow that providing a worker with mechanical equipment and power tools to improve individual productivity will benefit the economy as a whole. Typically, there are not enough trucks and power saws for all the drivers and construction workers who must transport and build. In such circumstances it is usually unwise for the entire economy to provide only a few workers with modern equipment and leave the rest with primitive tools. What is right for the economy as a whole depends on the size and quality of the labor force, the total availability of equipment and other resources, and the future and present goals of the economy.

This reasoning suggests that technical feasibility and labor and land productivity by themselves are not adequate criteria for choosing technologies. Rather, technological choices should be made in terms of general economic and social criteria. The economic evaluation of particular technologies is known as cost–benefit analysis. To perform such analysis, qualitative technical characteristics must be translated into estimates of how much of each type of resource is required by the technique. Then, by using wages and prices to value the costs of hiring and buying resources, the total costs of alternative production methods can be compared as well as the specific costs of using each resource. While such detailed comparisons will not be attempted in this report, the method will be implicit in all the economic analyses of the technological alternatives.

Describing technologies in terms of the amount of each resource they use makes it possible to distinguish two kinds of inefficiency. The common observation of technical backwardness in the developing areas has often been considered an indication of *physical inefficiency* in the use of resources. Strictly speaking, physical inefficiency exists whenever more of at least one resource is used than is absolutely necessary to achieve a particular amount of output. Assessment of *economic efficiency,* on the other hand, requires comparison of the total costs of using different sets of resources to achieve a particular level of output. A particular product may be yielded by a number of physically efficient techniques. But, of these alternatives, only the one that minimizes social costs will be economically efficient.

The proper choice of prices for evaluating economic efficiency is itself a complex issue. The "proper" prices should reflect the real scarcities of the resources in terms of the goals of the economy. These prices, called "shadow prices," would be generated by an economy that achieved its social goals with perfect efficiency. In fact, such economies do not exist. There are many sources of deviation from "perfection," some inherent in the nature of technology itself. For example, when "natural monopolies" result from economies of large-scale operations, market prices will not necessarily reflect the real productivity of the resources used. The essential operations of government in its taxing and spending may also interfere with efficient production. In addition, government powers are often used to support special-interest groups at the expense of the economy as a whole, with distorted prices the result. Because of the complexity of the problems of pricing resource inputs correctly, assessing economic efficiency is generally more difficult than assessing physical efficiency.[2]

Even in advanced countries, the "best-practice" technique typically differs substantially from the "average-practice" technique.[3] Yet the production inefficiencies of the developing countries are considered more profound and extensive. They have been associated with limited access to technical information and with a relatively poorly educated and trained labor force. Conventionally, such inefficiencies are attributed to the influence of traditional habits of thought and behavior on technical decisions as contrasted to the rational problem-solving approaches to management considered characteristic of industrialized countries.

The view that traditional production is inefficient has been challenged in recent years, especially with respect to peasant agriculture. It has been argued that peasant agriculture actually represents an optimal use of available resources. The means by which this optimality has been achieved have not been set out in detail, but the argument appears to be that a kind of Darwinian process of survival of the most efficient technologies has eliminated both physically and economically inefficient methods.[4] While this new view is plausible, the empirical evidence for peasant efficiency is equivocal at best, and the analysis and evidence for the existence of an effective selection process for efficient techniques is almost nonexistent.

A view still widely accepted and supported by some empirical studies is that important inefficiencies are commonly found in nonagricultural sectors of developing countries, both in the traditional "unorganized" sectors and in the modern "organized" plants. The frequent references to the necessity for greater literacy among workers and better engineering and organizational techniques among managers, taken in context, imply that there are neglected opportunities for improving output without substantial additions of physical resources. Yet the evidence is, perhaps by necessity, mainly anecdotal; it consists of isolated observations or, at best, a case study that is thorough but limited to a particular country at a particular time.[5]

Technological change means the replacement of economically efficient methods by still better methods. The replacement of steam by diesel locomotives on most of the world's railways and the replacement of open hearth by basic oxygen furnaces are dramatic examples. These innovations made it possible to produce more transport services and more steel with less equipment and manpower. As has been said, the effects of such technical change make it seem as if the laws of thermodynamics were repealed, for more output is achieved with given resources.

Studies beginning in the 1950's that tried to identify the sources of economic growth in Western Europe and the United States have suggested that technical change was responsible for most of the growth during this period.[6] These conclusions, as well as the obvious impact of new technologies in many areas of modern life, appeared to validate the post-World War II emphasis on research and development as the source of technological change and economic growth. The conclusions, which were also extended to Western Europe, were substantially revised in subsequent studies. Sources of growth other than simple capital accumulation and increased use of labor were diligently tracked down, and thus the role left for technical change was reduced.[7]

Yet technological change is, indeed, a remarkable phenomenon. It generates new products that relieve pain, cure sickness, and improve communication, and it increases output of all kinds. It creates problems of adjustment as well, and it has been argued that the problems it creates are more onerous than the problems it alleviates.[8] A complaint gaining currency is that the new technologies imported by the developing countries from the industrialized countries are not "appropriate" to their resource endowments and conditions. As a result, imported technologies do not help absorb their rapidly growing labor forces and make demands for materials, organization, and skilled personnel that cannot easily be met. It is implied that appropriate technologies can provide more of the benefits of technical change with fewer deleterious side effects.[9] The essential questions are the availability of specific alternative production methods and their economic and social costs and benefits. These issues cannot be settled by assertion. They will be the continuing focus of analysis in this report.

It has also been argued that the agents of technical change—the innovators and entrepreneurs—are the central figures of the development process. It is they who implement the new methods and introduce the products that provide the major new investment opportunities, and these in turn result in the improved worker productivity and the production of new goods that are the essence of economic growth. The success of these innovators encourages others, more timid, less qualified, or less opportunely placed, to imitate the specific innovations and entrepreneurial patterns of economic behavior.[10] By this reasoning, the process as well as the content of technological change is significant for development.

Capital Accumulation and Saving

Outside their modern production enclaves, developing countries have limited tools, equipment, and other capital facilities. The standard examples that come to mind are farmers with simple stick plows, construction workers using picks, hoes, and baskets to move dirt and rock, and artisans using mostly hand tools and little, if any, powered equipment. While these examples do not accurately characterize all the production activities of any developing country, they do reflect in extreme form the relative capital scarcity common to all the developing countries.

When capital is so scarce, it is not surprising that labor productivity and income are low and that poverty is widespread. The obvious remedy is more capital, and that is certainly a necessary condition for development. But through emphasis and/or neglect of other conditions, capital investment has often been treated as a sufficient condition.[11] That is certainly wrong, since a great deal depends on the effectiveness with which capital is used and that, in turn, depends on many other features of an economy.

The theories of growth developed in the 1950's, which related increases in income and output to increases in capital created by new investment, were attractive in their simplicity and their apparent power both to explain the existing poverty and to plan for future development. Though the investment–growth theories often neglected the role of labor, that did not seem to be a fundamental omission when they were first applied to the "labor surplus" economies of south Asia. Even now a simple relation between increases in capital and associated increases in output is often the basis of much development planning. When used skillfully and supplemented by consideration of manpower and other requirements for growth, the capital–output relation can be a powerful analytical and policy tool.

To increase the capital stock by net investment there must be saving. When resources are fully employed, they must be saved from current consumption to make them available for capital formation. Saving can be generated from several sources. Saving is perhaps most often thought of in private and personal terms: individuals, by not spending all their income on consumption goods and, thus, by not exercising all their claims on current output, make some of it available for new capital formation. Private business saving is at least as important as personal saving in many of the developing countries. Businesses, by not distributing to their owners all their net profits, also effectively restrict consumption and, thus, free resources for investment in new capital. Governments and government enterprises also generate saving by restricting their expenditures on currently produced and consumed services to less than their tax income or revenues. Typically, any surpluses of this type are moved directly into capital formation by government, but they could be made available, directly or indirectly, for private capital formation. Finally, foreign resources used for capital formation can also be considered "saving"—foreign saving—and the basis of new investment.

One advance in the economic understanding of development processes has been the recognition of the limits to the potential substitutability of savings from domestic and from foreign sources. Specific types of facilities and equipment are necessary to create each specific type of investment goods. If those facilities exist domestically, then domestic saving can help make them available for new capital formation. However, if the necessary production facilities are not available domestically, no amount of domestic saving will directly free resources for new investment. For example, a developing country that wants to invest in a hydroelectric station of the largest size must build dam and powerhouse structures of concrete and install large turbogenerators. If there are domestic cement plants, the cement for the concrete will be released from domestic supply—if personal saving limits private housing demand, for example. But only a few countries in the world have the specialized metal-making and metal-working equipment required to produce the largest turbogenerators. So those must be imported.

However, technically necessary imports of investment goods can be made available indirectly via domestic saving if such saving makes possible export of goods that would otherwise not be exported. Thus, domestic saving, which reduces domestic purchases of textiles and makes them available for export, in effect earns foreign exchange to pay for imports of turbogenerators as well as other types of goods. Yet there are limits to the rate at which exports can be expanded, and these in turn create limits to the indirect substitution of domestic for foreign saving for most developing countries.[12]

According to one view of economic growth processes, there are limiting factors to what could be achieved by capital accumulation. Diminishing returns to new investment, like diminishing returns to labor in the Malthusian view, would steadily reduce the net contribution of new capital formation. A more recent view, embodied in most of the development planning models, holds that the right kind of investment can break nearly all production bottlenecks. In this view, almost all "fixed factors," other than geographic conditions, can be expanded by investment without diminishing the returns to the investment. Even the availability of arable land can be increased by irrigation and other land-improvement schemes.[13] However, there will never be any definitive tests of the alternative viewpoints, partly because of another essential aspect of capital formation: its function as a carrier of new technology.

Although difficult to measure precisely, an important part of most new investment carries technical change with it. Standard types of investment goods can be bought "off the shelf," but much new investment is "tailor made," and the tailoring process typically adds new technical arrangements, devices, and other improvements even without major innovations. Thus, capital formation and technological change are not separate growth processes, though it is often convenient to discuss each separately.

Growth and Improvement in Quality of the Labor Force

One of the important differences among the developing countries is the relative population densities and the ratios of available labor to capital and other productive resources. There are "overpopulated" and "underpopulated" developing countries, where these terms reflect general impressions of differences in labor/resource ratios. These differences make generalizations difficult, but some features of population and labor-force issues are common to most developing countries:

- Rapid population growth rates;
- Substantial and growing open unemployment;
- Large-scale rural–urban migration; and
- Education and skill patterns in which significant shortages at some levels may coexist with surpluses at other levels.

With only few exceptions, developing countries have high rates of population growth. These rates are nearly always substantially over 2 percent per year and often 3 percent per year or more, whereas the growth rates in most advanced countries are 1 percent or less. The relation of the labor force and its growth to total population and its growth depends on the age structure of the population and participation in the labor force. A relatively high proportion of a rapidly growing population is too young to be in the labor force. In addition, for reasons of culture and religion but also because of preoccupation with child rearing, the participation of women in the labor force, at least in urban economic activity, is often relatively low in the developing countries. Nonetheless, because the relatively high overall population growth rates have been sustained for some time in most of the developing countries, the growth rate in their labor forces is also now relatively high.

The projected rapid increase in the labor forces of the developing countries compared with current labor-absorption rates has made unemployment one of the dominant issues in the developing countries. The current unemployment figures, partly because of their statistical unreliability, do not now indicate a widespread and overwhelming problem. However, urban unemployment in many developing countries is growing rapidly and is already considered a major issue. The growth statistics and expectations, combined with the information now available on the growth of employment that has accompanied development over the past 30 years or so, produce a gloomy forecast. A calculation by Turnham is typical:

By simple arithmetic, a manufacturing sector employing 20 percent of the labour force would need to increase employment by 15 percent per annum if only to absorb the *increase* in a total labour force growing at 3 percent per annum. In addition, because of productivity increase, about 3 percent growth per annum in output seems needed to maintain a constant labour force in manufacturing.[14]

The potential for expanding employment depends on capital accumulation rates, the labor force/resource ratios prevailing in these sectors, and the degree to which the new technologies permit additional labor to be used with additional resources. For example, some developing countries, though capital-poor, are rich in mineral resources and in underutilized agricultural land. These resource-rich developing countries have "frontiers" to exploit and, as claimed for the United States during its most rapid population growth in the nineteenth century, these frontiers may provide a "safety valve" for the absorption of labor. However, even in the relatively resource-rich and "underpopulated" developing countries, difficult unemployment problems may emerge as they adjust to rapid population and labor force growth rates and to the new patterns of sectoral growth. In developing countries not richly endowed with agricultural land and/or important mineral deposits, urban manufacturing and the service sectors must absorb most of the increases. That, in turn, requires new investment to provide the equipment necessary for the productive use of labor. So labor absorption depends on investment rates as well as the degree to which the technologies employed permit intensive labor use.

Although overall growth in population and the labor force is one of the major factors aggravating unemployment, rural–urban migration is also a contributing factor. These migration processes are not completely understood, but rural–urban income differentials apparently play an important role. These differentials may arise from real productivity differences in rural and urban areas that, in turn, depend on the use of modern capital-intensive technologies in urban production. The income differentials may be created to induce higher productivity by reducing labor turnover and/or by improving nutrition and providing other social amenities associated with higher income. But the rural–urban income differentials may also be the result of minimum-wage laws or union organizing, which is effective in urban areas.[15]

Typically, there are significant shortages of skilled labor in developing countries even when unemployment is substantial. While development based on natural-resource exploitation is the easiest in many respects, even that requires new processes. And, if development is to spread and become permanent, manufacturing and construction sectors must be expanded and new processes must be adopted. In the agricultural sector, development typically requires new seeds, fertilizers, pesticides, and, often, changes in irrigation patterns. All of these changes require new and additional skills in the labor force.

The creation of labor skills through education and training is now conventionally considered to be a kind of capital formation—"human capital" formation. Time and resources are required for education and training, just as they are in the formation of physical capital. And education and training can increase the return to labor, just as investment increases the potential productivity of resources. To the extent that on-the-job or in-the-field training also creates skills, it is also human capital formation. However, while there are insightful analogies between physical and human capital formation, there are important differences as well, in the inalienability of human labor and in the individual and social character of the decision processes that determine its use. These differences require caution in the use of the human capital–physical capital analogies.[16]

The developing nations vary widely in their stock of human capital and in their facilities for creating additional amounts of such capital. In some countries, formal education is an old tradition, so that it is a familiar, if not universal, social institution. In other countries, the experience with formal education systems is relatively new. In some countries, artisan and household enterprises and small factory establishments serve as important training grounds for workers. In other countries, this type of undertaking is relatively unimportant. Also, depending on their culture, language, and relative proximity, some developing nations can more easily and extensively use skilled workers from other countries. Developing countries that can produce highly qualified persons, but cannot compete in the real income offered to these people, can be faced with "brain drain." This "export" of talent may or may not be beneficial, depending on the returns that accrue.

That new technologies often require new labor skills is clear, but there is so little evidence on the relations between capital intensity and labor-skill requirements that no generalizations seem warranted. In some cases, for example, modern and capital-intensive equipment simplifies the labor operations by specializing the tasks, or it carries automatic control devices that reduce the need for human skill in regulating the equipment. Adoption of new technologies will almost certainly change the composition of workers employed. In some situations, only a few highly skilled workers may be necessary, and the rest of the labor force can be relatively untrained. Thus, the education and training of labor required by new technologies is not readily evaluated, and few studies are available.

Changes in Sectoral Composition

Development in every country certainly means some degree of industrialization. It also means expansion of the power-generation sector, the transportation and communication sectors, the construction sector, and all the marketing and financial service sectors. It requires growth in public services of all types: education, public health, administration, and so on. Some patterns of sectoral change are more conducive to rapid growth than others, especially when recourse to foreign trade is limited. In general, rapid economic growth is associated with concentration of a relatively large proportion of investable resources in the capital–goods sectors at the early stages of the growth process. Yet this generalization is subject to so many qualifications that it does not provide a useful general guide to policy. Each country requires its own separate analysis to establish its own optimal growth path. For example, small countries cannot establish efficient capital–goods sectors as readily as large countries, because economies of scale are important in industries that require relatively large levels of output to be efficient. The emphasis in development policy on the different economic sectors implies an emphasis on technologies having different implications for employment opportunities as well as for overall economic growth and income distribution. It is understandable that the relative growth of the various sectors has been one of the most controversial issues in development policy.

In addition, sectoral interdependence arises because flows of inputs and outputs among sectors are necessary if each sector is to deliver the final goods and services demanded of it. To invest in fertilizer plants, cement and steel have to be delivered to construction projects. To produce cotton textiles, fertilizer must be delivered to farms and used with labor and other resources to grow the cotton, which must then be delivered to textile factories, and so on. It can be intuitively appreciated that the interdependence of the sectors requires some kind of balance to prevent major bottlenecks in the economy. Factories constructed to use centrally generated electric power will stand idle if the power-generating stations are not ready when the factories are completed. That idleness means that the investment will pay off more slowly and less well than it should have, whereas coordination of factory and power-station construction would avoid such idleness and waste.

Thus, sectoral interdependence makes it impossible to judge a technology solely in terms of its direct effects on output, employment, and so on. Because nearly every productive activity both sends and receives goods and services from other productive activities, its indirect consequences must also be taken into account in a comprehensive evaluation.

One of the major development-policy controversies regarding sectoral balance has concerned the relative emphasis on production of goods intended to substitute for imports or of goods intended mainly for export. Import substitution, it has been claimed, would reduce the developing countries' need to export primary products at increasingly unfavorable terms of trade to purchase the manufactured goods necessary for their development. Against this policy, it is argued that because domestic markets for many manufactured goods in developing countries are relatively small, import substitution has actually led to the establishment of uneconomically small-scale plants.[17] In addition, it is claimed that protection has permitted developing countries to use technologies that do not take full advantage of their particular labor and other resource supplies.[18]

On the other hand, it is claimed that export promotion would exploit the comparative advantage of the developing countries in their special natural resources and, often, relatively inexpensive labor. Exposing the economies of the developing countries to international competition would both permit them to buy needed goods at relatively low prices and stimulate internal efficiency in the use of resources. Also, foreign trade is a many-purpose balancing sector. Because foreign trade can provide many kinds of inputs not domestically available and take many kinds of outputs not domestically absorbable, it can make possible large-scale and sometimes less costly production. Foreign trade can also reduce, though not eliminate, the need for input–output balances among sectors within a country.[19]

But the sectoral-balance controversy goes beyond the import substitution–export promotion issues. Special emphasis on industrial expansion—and within industry on the expansion of heavy manufacturing industry—as the key to development progress was an early part of development ideology. Concentration on investment-goods production necessarily implies relatively high savings rates and lower consumption rates. As a result, however, the potential levels of consumption in the future would be higher. Thus, the sectoral emphasis on investment goods or consumption goods also involves a decision about the distribution of the benefits of development over time.

There also appear to be characteristic differences among sectors in the opportunities for the efficient and intensive use of labor.[20] If electric power is to be generated in a central station, it is difficult to find efficient, labor-intensive methods. Likewise, there are no known ways of producing steel efficiently except by using a lot of plant and equipment. In both cases, there may be more opportunities for intensive use of labor in the peripheral processes such as fuel and other material transport than there are in the core processes.[21] However, that these have a major effect on employment opportunities is yet to be shown. On the other hand, other manufacturing sectors (such as in some consumer goods, and certainly in agriculture, the services, and construction sectors) appear to permit more opportunities to absorb labor efficiently. These possibilities will be examined carefully below.

Agriculture, which was relatively neglected in the early stages of the development policy of a number of countries, has more recently come to be widely regarded as a premier sector. In agriculture, the technological choice issues have focused on the emphasis that should be given to methods that rely heavily on mechanical equipment and chemical fertilizers and pesticides. These questions have been related to agricultural organization, the concentration of land ownership, and the potential displacement of labor as a result of new technologies.

Changes in Economic Organization

Changes in economic organization are often considered one of the major aspects and conditions of development. Some of the envisaged changes have to do with the relative role of private markets and government ownership and direction in the economy. But development also involves changes in such fundamental social patterns as the role of the individual household enterprise as compared with wage-paying capitalist or cooperative types of enterprise. While it is difficult to evaluate the significance of such changes, they cannot be ignored as an essential part of the development process.

Rightly or wrongly, the lack of development in many countries has been considered a failure of private capitalism. This reasoning is due partly to the association of colonialism with the capitalism of the colonizing country. In countries without a colonial history, the stagnation of the past has also occurred when the dominant economic organization was a form of private capitalism. While historical relations are often far from conclusive, many of the developing countries have reacted against private capitalism as part of a widespread ideological trend. However, the government's role in the economies of the developing countries has been related not only to their economic ideology but also to their political development. Government intervention in economic decision making and control has enabled them to penetrate the social systems and institutionalize authority when action with other instruments and in other areas has been difficult to sustain.

The replacement of the private organization of production with public or quasi-public organizations has to some degree been a feature of the development process in many countries. Nationalization of enterprise has been due partly to economic ideology, but partly also to the political goals to which it contributes. Typically, the major "utility" sectors of the economy, such as transportation and power, have been nationalized and, in some countries, important parts of manufacturing have been nationalized as well. In some cases, the nationalized sectors have operated directly under government ministries, but, in other cases, new government corporations have been created.

Government operation of production enterprise, directly or indirectly, has encountered many problems. Staffing patterns and organizations different from the civil service have been difficult to establish successfully. The multiple objectives typically set for government or quasi-public firms have created special difficulties in the developing countries. A private firm is expected to maximize its profits within the technological and demand constraints of its sector and the taxes and controls of the government. However, public enterprises are often subject to special constraints on the pricing of their product, the level of their wages, and their employment conditions. The degree to which public enterprises can adjust to their goals and constraints depends in part on the nature of the technologies they employ.

Many of the developing countries have turned to government planning and government controls for the direction of investment allocation. Review and control processes in the economic ministries may both direct government enterprise and regulate private enterprise. The level of detail at which the centralized control operates varies considerably among countries. Often only the sectoral composition of new investment is controlled, and the choice of technological methods is left to individual enterprise. But in some cases the latter is also centrally regulated.

Other countries rely heavily on various market mechanisms to determine the sectoral allocation of investment. Under these conditions, the financial markets have a central role in development decisions; they affect investment allocations and the choice of technology, and they determine which individuals or groups will undertake particular investments.

In addition to overall policies for economic organization, there may be special sectoral policies, as in the formation of farm cooperatives, for example. Such cooperatives may be encouraged, when individual farms are relatively small, to facilitate the use of technologies that are most efficient if used on a large scale. Or cooperative organizations may be used to achieve other advantages of large scale in purchasing and marketing.

Finally, systematic changes occur in the most prevalent types of economic organizations in the course of development, even when there are no explicit government policies toward enterprise organization. In the developing countries, much production is typically organized in small-scale, often family, enterprise, not only in the agricultural and service sectors but also in the early stages of manufacturing. This type of enterprise is essentially "noncapitalist" in its use of labor and may be relatively labor intensive since family labor need not be paid standard wages. Thus, the technologies chosen and used in family enterprise can, in general, be expected to be different from those adopted by enterprises that hire and pay for labor through more explicit wage contracts.[22]

Although family enterprise persists over the course of development, it tends to become less important in one sector after another. Its changing importance is only one example of the spread of market-determined resource decisions in the course of development. For example, in some developing countries agricultural land is not bought and sold or rented in markets where it is valued only for its productivity. But these conditions change over time. This, too, has important implications for technologies, since absence of market-determined resource valuations will lead to different land–labor–capital combinations than those determined by markets.

Social and Political Development

Economic development and the accompanying political changes are an aspect of the modernization of the developing nations. The term "modernization" is not intended as a pejorative characterization of the "traditional" societies from which the developing countries are evolving. Rather, modernization is intended to refer to the whole complex of changes, economic and otherwise, which occur in the course of development. These include the differentiation of institutional structures to perform societal functions and changes from personalistic to more formalized structures and procedures of social organization and decision making. The economic changes are, perhaps, the most obvious and are most obviously associated with technological decisions, but the effects of modernization are pervasive and are intimately associated with the production technologies used.

Some of the clearest and most dramatic examples of the relations between social structure and technology can be found in economies that still rely on hunting to supplement small-scale agriculture. In these societies, hunting is typically a male activity, and status as a male and among males depends on success in this activity. Substituting other food for game is a technological innovation that destroys a status-creating activity and can impose severe social tensions on such a society.[23] While the social effects of technological changes in other societies are often not so obvious, they may be just as profound. Such change, however, is not necessarily destructive in all respects. For example, it is commonly argued that in the early phases of Japanese industrialization the conversion of the samurai warrior class to industrial entrepreneurs and managers was both cause and effect of the technological changes transforming that economy.[25]

The changes in the distribution of work and production responsibilities among family members that accompany modernization in turn change the family structure itself. One of the major concerns of the families of traditional societies is organization for production. This responsibility is transformed with new ways of production so that family structure itself undergoes major alterations.

Modernization also changes the way people associate themselves in larger communities and particularly their politicization. The political development accompanying modernization is at least indirectly related to the technological changes that are a fundamental characteristic of modernization. Specific sectoral emphases and technological innovation may promote types of political developments that strengthen the capacity of governments to interact with their citizens. For example, concentration on large-scale industry is more consonant with centralization of political authority than decentralized development based on rural industry. In turn, political development can contribute to economic development by strengthening the capacity of governments to identify and deal effectively with economic problems. Even in the most private enterprise and capitalistic economies, government has important economic functions to perform that will determine, to a considerable extent, the effectiveness of the private sector. When government participates substantially in the economy through ownership, control of enterprise, and/or planning and overall direction, the government's ability to carry out its responsibilities is even more critical to development.[25]

REFERENCES AND NOTES

1. For a somewhat deeper but still relatively accessible survey of development processes see J. Bhagwati, *The Economics of Underdeveloped Countries* (1970).
2. The nature of the distortions that require the estimation of shadow prices are discussed in P. Dasgupta *et al.* (1972) and I. Little and J. Mirrlees (1969).
3. H. Leibenstein (1966).
4. T. Schultz (1964), Chap. 3.
5. Louis Wells, "Economic Man and Engineering Man," *in* C. Timmer *et al.* (1975).
6. R. Solow (1957).
7. E. Denison (1962), (1967) and Z. Griliches and D. Jorgenson (1967).
8. For a critical statement on the problems created by technological change—and economic growth—see, for instance, Mishan (1970).
9. E. Schumacher (1973), Part II, Chap. 5.
10. E. Hagen (1962)
11. There are few if any analysts who have explicitly taken this position. But it has emerged from the central place given to investment policy in most development analyses and, more to the point, in policy actions in developing countries.
12. These insights into the differential requirements for domestic and foreign saving

required for capital formation have entered the development literature primarily under the headline of "two-gap" analysis. One gap is the possible difference between the domestic saving and the investment desired for the development program. The other gap is the difference between foreign-exchange resources and import requirements. Of course, actual savings from all sources must finally match the actual investment actually carried out, and the two gaps must both disappear. Ex post, see H. Chenery and M. Bruno (1962).

13. For a comprehensive view of development planning models and their embodied assumptions see C. Blitzer *et al.* (1975).
14. D. Turnham (1971), p. 10.
15. J. Stiglitz (1972) and J. Harris and M. Todaro (1970).
16. See, for example, R. Eckaus (1962).
17. See I. Little *et al.* (1970).
18. See, for example, G. Ranis (1974).
19. For the arguments for an "export-led" strategy of development see, for example, I. Little *et al.* (1970).
20. These characteristic differences are at the heart of the modeling of production in dualistic economies *in* A. Kelly *et al.* (1972).
21. See G. Ranis (1974).
22. For examples of arguments along these lines see A. Sen (1975), Part III.
23. For examples of such changes see E. Service (1962).
24. Y. Horie (1965).
25. N. Choucri (1976).

3 Alternative Criteria of Appropriateness of Technological Decisions

Introduction

Since technological decisions affect the patterns of all development processes, they may to some extent be consciously used as instruments of policy to affect the course of development. Recognition of technology's essential role and its potential as a policy instrument has been implicit in the recently burgeoning interest in implementing "appropriate" technologies. This chapter will consider the specific development goals to which technological choices may contribute and the degree to which these goals are complementary or competing.

The only standard for deciding the "appropriateness" of technological decisions is reference to the general goals of development. The qualities of technology are not more or less desirable in themselves but only for their output potential, their corresponding input requirements, and their effects on social and political organization. In particular, small-scale or labor-intensive technologies are not necessarily "appropriate" because they are small scale or labor intensive. Whether they are appropriate depends on their ability to contribute to development objectives. Only if inherent features of technology dictate other patterns of life and development can technology be treated as an end rather than as a means. While this possibility must be considered, the resolution of the issues involved should not be prejudged by adopting this view. The alternative viewpoint is that, although technological choices have broad economic and social implications, a technology seldom has inevitable consequences that dictate a particular pattern for the economy and society. Many economic and political instruments can, in principle, be used to guide and modify the impact of technical decisions. However, this viewpoint also requires critical examination, since there may be many circumstances in which there are limits on the extent to which the undesired effects of a particular technological decision can be ameliorated.

The major goals enunciated for development will be examined in this chapter and interpreted as criteria for appropriateness of technological decisions. Essentially, these goals and criteria are related not only to the quantities of income and output generated over time but also to how they are produced and distributed. As will become clear, there is some redundancy in these goals, and even opposition. The analysis will not be organized to avoid these features but rather to allow them to emerge in the course of the discussion.

Maximization of Net National Output and Income

Perhaps the most conventional formulation of the national economic objective of "doing as well as possible" is the maximization of current national output and income. Yet the goal should not be understood in a static sense to mean doing as well as possible only in the current year. Rather, it nearly always implies maximizing the discounted sum of future outputs over some time horizon or, alternatively, maximizing output in some future target year.[1] This objective is attractive as a primary goal of development because, in general, it provides the largest sized "pie" to be divided among various uses and allocated to various groups. There can be little doubt of the support for this general goal: the desire for more of the material things of life—food, clothing, shelter, and some "luxuries" beyond that—is evident in the intensity with which economic objectives are pursued. And suppose the goal were not being achieved. Then it would be potentially possible to increase the amounts of goods and services available to everyone, or at least to some groups and uses, without decreasing the availability of goods and services in any other direction. Thus, if the goal were abjured, the potential would not be pursued, in spite of obvious needs.

The rules for technical choice to maximize the national output and income at any time and over time require physical and economic efficiency in each productive establishment and in the economy as a whole. The specific content of these rules has been worked over intensively by economists. They are embodied in the methods of cost–benefit analysis that can be applied to new projects and technical decisions and in the normative overall planning techniques that have been developed. In general, neither efficiency within any enterprise nor overall efficiency is achieved by maximizing the productivity of any single resource in any particular use. Rather, efficient use of resources is achieved by equalizing the incremental returns to equal expenditures on each type of resource. If such returns are equalized, then no reallocation of resources could possibly increase output.

Concentration on national output and income maximization as a development goal has significant limitations, however. It avoids the essential questions of what should be the composition and distribution of national output, and it implicitly assumes that the national output can be used directly or indirectly for any social purpose and distributed according to any social rule. Yet these assumptions are not warranted. For example, investment goods are seldom useful for consumption purposes, products for export may well be different from those consumed domestically, and political constraints may limit the scope of income redistribution that is feasible. Thus, development goals increasingly are cast in terms of more fundamental national objectives such as achieving particular levels of consumption, patterns of income distribution, self-sufficiency, and so on.

However, even when the overall goal is not to maximize output and income, *and* when the use of economic policy instruments is constrained, *and* when market or other allocation mechanisms are not fully effective, efficient use of resources within each production establishment may still be desirable.[2] So caution in the sacrifice of such efficiency is always necessary.

Maximization of Availability of Consumption Goods

Since most would agree that consumption in some form is the final goal of economic activity, and since consumption goods are not perfectly substitutable with other goods, it is reasonable to set maximizing consumption as the goal for the developing countries. This goal is in fact enshrined in many development plans and planning models as well as in one major approach to cost–benefit analysis.[3]

However, the goal should not be interpreted in the narrow sense of creating consumer enjoyment in the short run. Instead, the maximization is intended to occur over some foreseeable future period. Thus it is necessary to consider how present decisions about technology will affect the production of consumption in the future as well as in the present. Those effects in turn depend on investment and savings rates. So the choice of appropriate technology (that is, one that would maximize the attainment of the consumption availability goal), must consider any possible differential effects on saving. The differential effects may arise because alternative technologies imply different proportions of wages, capital income, and tax payments in total income. Since a future loaf of bread cannot satisfy current hunger, the choices must also weigh the relative importance of present and future consumption. The weighing of present and future must also reflect foreign-trade constraints, which affect the viability of any particular time stream of consumption.

Again, however, this goal is stated in terms of a national aggregate. It implicitly assumes no need to worry either about the composition or the distribution of consumption, perhaps because taxes and subsidies can be used for particular distributional objectives. When this is not the case, more specific development goals must be set.

Maximization of Rate of Economic Growth

One of the most frequently articulated goals of the developing countries is to increase their rate of growth or, in a more extreme version, to grow as rapidly as possible—that is, to *maximize* their rate of growth. This objective seems plausible until it is realized that growth requires current sacrifices, and maximizing growth means maximizing those sacrifices. The critical sacrifice is current consumption because, as pointed out in Chapter 2, consumption must be restrained to free resources for investment to add to production capacity to create more investment, etc.

A more acceptable version of the objective is to maximize growth subject to the maintenance of a minimal and, perhaps, rising level of income or consumption. In this form, the goal becomes a version of that articulated just previously: maximizing a combination of present and future consumption, since the object of growth is to increase consumption possibilities in the future.

In effect, this criterion of appropriateness of technological decisions may only imply a different weighting of present and future output or consumption rather than an objective completely different from the previous two. But the significance of the different weights for the composition of output and the choice of technology can be profound. An emphasis on growth will place stress on technologies that help generate a high rate of saving for investment. An emphasis on growth also tends to be associated with relatively high interest rates, which, in turn, works against capital-intensive technologies and for methods that pay off rapidly.

Reduction of Unemployment

As noted, the problems generating the greatest concern about the appropriateness of technological choices have been employment and unemployment. Despite the substantial amount of investment that has occurred in most of the developing countries, their unemployment problem has not been relieved. Indeed, partly because their population growth rates and their labor force growth rates are high compared with their growth rate of employment, unemployment is expected to increase. It has been easy, though not necessarily correct, to attribute the low growth rate of employment to the choice of "inappropriate" technologies, either locally generated or imported from abroad. The next step is usually a call for more labor-using technologies. However, before taking this step it is desirable to consider the social functions of employment in more depth.

First, employment in conjunction with other types of productive resources generates output. To perform this function as well as possible, the appropriate choice of technology should be that which maximizes output, taking into account the type of output desired and the preference for present or future increases in consumption.

Second, employment is a means of distributing income. In effect, labor is made a condition of the receipt of income. The amount of income is associated with work time and usually with the quality of the labor as well, where quality reflects the work's contribution to the output's value. While the income-distribution function of employment is not always separated from its output function, both in principle and in actuality the distinction can be and is actively made, as, for example, in welfare assistance and old-age-support programs.

Social recognition and personal satisfaction are other features of employment: the recognition aspect.[4] Social status and self-evaluation are a function of the kind of work people perform, not just in relatively traditional societies but in modern systems as well. The individual trauma associated with involuntary unemployment in the advanced Western economies is an example of the relation of employment and status. Typically, the socially approved roles for individuals permit only small and specially identified classes to be more or less permanently absent from formal or informal employment without loss of status and self-esteem. In addition, the redistribution of income through government programs may not be a realistic option, given the limited capacities of the government to mobilize resources and distribute them to potential recipients. It is partly for these reasons that distribution of income through employment, even with economically inefficient technologies, may be socially acceptable, while distribution through dole or welfare schemes is less so. In addition, if there is obvious inefficiency in terms of the output or growth criteria, any burden or blame can be shifted to less easily identifiable bureaucrats and managers.

Finally, the reverse of the income distribution function of employment is its "tax" function. Employment requires the sacrifice of leisure and, usually, the personal direction of each individual's own activities. If the income received from employment were totally within each worker's disposal, the burden would be completely self-determined. But taxes and rules for income subsidies make employment to some degree a socially determined tax.

Because employment performs several different functions in every society, there is no single employment or unemployment criterion for the appropriateness of technological choices. In fact, there may well be conflicts among employment criteria. It has been argued that when technologies considered appropriate by income-distribution standards lead to inefficient choices by output maximization or growth standards, the latter criteria of appropriateness should prevail. This would result in the largest output to distribute. Conceivably, then, any desired pattern of income distribution might be achieved by one or another redistribution scheme. In principle, everyone would then be better off than if output, in the short or long run, were sacrificed for a particular distributional goal. Yet income redistribution in the patterns desired may not only be beyond the capacities of government agencies in developing countries but may not be as socially acceptable as a government subsidy. By comparison, a particular income-distribution pattern achieved by the conscious adoption of inefficient technologies may be more readily tolerated.

The sacrifices of output in the choice of technology to achieve distributional goals will often be difficult to identify. When distributional goals are achieved through technologies that use more labor than is strictly necessary, the distinction between persons "on the dole" or "on welfare" and those "righteously working" disappears. Technologies that increase employment may also go far toward meeting the standards of appropriateness by the criteria of social recognition and personal satisfaction as well as income-distribution criteria. Employment-based criticisms of the technological choices made in the developing countries tend to emphasize the distributional and self-esteem functions performed by employment rather than the production and tax aspects. What is not clear in many of the criticisms, however, is the extent to which these latter functions of employment should be sacrificed for the former goals, nor is there even a recognition that a trade-off may be unavoidable.[5]

Redistribution of Income and Wealth

There has been increasing concern in recent years about the distributional aspects of development and a growing interest in policy instruments that can affect these aspects. It is natural, therefore, that technological decisions should be given attention in this search. Yet any policy movement in this direction is strongly constrained by lack of general understanding of the factors determining the personal distribution of income and, in particular, the effects of technological decisions. For example, there is some evidence that the distribution of income in the developing countries is more unequal than in advanced countries but no firm comprehension of why this might be so.[6] However, it has been noted that appropriate technological decisions can affect the distribution of income by affecting the volume of employment provided. These decisions may have potentially similar effects in the quality of employment created, depending on whether the employment is mainly of skilled or unskilled workers.

Distributional issues also arise in relation to the precise meaning given to the consumption criteria for technological choice, since potential additions to the consumption of high-income groups generally are not given the same social valuation as additions to the consumption of low-income groups. The benefits of alternative technological choices would, therefore, be weighted differently, depending on whether they give employment and thus generate income mainly to low- or to high-income groups.

The goals of maximizing aggregate or average per capita consumption over some foreseeable period on the one hand, and that of spreading consumption to specially benefit low-income groups on the other hand, may conflict because of different savings levels associated with technologies chosen by the different criteria.

Regional Development

Regional development goals are simply economic, political, and other social goals specified for a particular locality. If there is any difference between such regionally specified goals and those set forth at the national level, it is that the regional goals may be described in terms relative to the average of the nation as a whole or some particular "advanced" region. For example, the growth goals for the less-developed south Italian Mezzogiorno have been specified relative to the highly industrialized Italian north.

The reasons for assigning special priority to particular regions are fundamentally political, but they may emanate from an appreciation of particularly profound differences in economic conditions and the degree of participation in overall national development, as appears to be the case with the Brazilian northeast region. However, regional development goals may also be frankly political and involved with conflicting international territorial claims, as in the case of the extreme north of Chile vis-à-vis Bolivia. Or the politics of regional development may be completely internal and related to differential support to particular factions, parties, or ideologies.

Differential rates of regional development are responses to differential availability and use of resources, including regional differences in the level of investment. Production methods are likely to be less important than investment in particular types of production and location decisions. However, because regional potentials differ, a decision to emphasize development in a particular area will result in emphasis on particular sectors and processes. For example, the decision to encourage the development of an agricultural region characterized by inadequate water supply will give special weight to all the investment and technical decisions related to water use. If a water supply technology requiring relatively large-scale investment by individual farmers is adopted, then the large farmers will be favored over small farmers, unless cooperative organizations and special financial programs for small farmers are sponsored. Similar considerations also apply when technological decisions are not guided primarily by regional objectives, but when they are related to the explicit and often large-scale problems of regional development the implications may become more obvious.

Regional development is also affected strongly by regional availability of natural resources. The great variability in these will typically dictate local adaptations of technology: in agriculture, mining, forestry, and so on. In resource-based production, the local resources greatly affect the choice of technologies appropriate by the economic criterion of cost minimization or output maximization. However, in resource-based production, there is no greater understanding of technologies appropriate by criteria other than output maximization than there is in other types of production.

Balance of Payments Relief

There are two quite opposite theoretical and policy arguments in which technological decisions are associated with balance of payments policy. One argument claims that concentration on export promotion policies, i.e., investment in industries that expand export sales and earnings, will cope best with the balance of payments problems that arise in the course of development. The opposite policy would place more emphasis on industries whose output substitutes for imports. While the arguments are mainly about sectoral investment allocations, it has also been argued that the different sectoral investment policies employ different technologies as well.[7]

The argument for export promotion, briefly, is that for export industries to thrive in the competitive world environment they must reflect the real comparative advantages of the developing country. This, in turn, can be achieved only if the technologies used in the export industries make optimal use of available resources to minimize costs. Thus, the criterion for the identification of appropriateness in this connection is really the one first mentioned above: output maximization or cost minimization with available resources for any particular level of output produced.

The argument for investment in import-substituting industries has several themes. One is that the terms of trade are steadily worsening for the typical primary commodity exports of developing countries and that the way to avoid the consequences of this trend is to create domestic industries that will produce the imported goods that had been paid for with commodity exports. However, these industries were also intended to use resources efficiently in the sense that output was maximized or costs minimized for any particular output. Another theme in the import-substitution argument is that the policy will make developing countries more independent and self-sufficient economically, and that end is desirable in itself.

Thus, both the export-promotion and import-substitution policies are focused on output maximization but differ in the means chosen, except that the latter may also be justified on nationalistic and autarkic grounds.

However, it has also been argued that actual pursuit of the export-promotion policy has led to the choice of efficient and appropriate technologies, while this has not been the case when import-substitution programs are followed.[8] Although import-substituting industries were not intended to be inefficient, that often appears to have been the outcome because the domestic markets are small in most of the developing countries. Moreover, it has been argued that government interferences with, and distortions of, production and pricing are more pervasive when import-substitution policies are followed, because domestic production can be "protected" while exports are exposed to international competition. Because of this competition, pressure to find technologies that minimize costs will be greater in production for export than production for import substitution.

It is well-known that government can use subsidies and other favored treatment to influence decisions in export sectors just as much as in import-substituting sectors. The claim that governments have foregone the opportunities to exercise distorting influences on exports remains unsubstantiated. However, the issues are controversial ones, and the arguments over these two policies emphasize the need for cautious judgments. Technological decisions and balance of payments policy are also related by the foreign-exchange cost of importing technologies from abroad. Some studies have indicated that license fees and patent royalties involved in importing foreign technologies can become a significant part of foreign-exchange requirements.[9] So domestically developed technologies or foreign technologies acquired at relatively low cost would be the more appropriate technologies by this test alone.

To increase foreign-exchange earnings, some developing countries that export raw materials insist on increased domestic processing of these exports and have raised this policy to the status of a development goal. Yet raw-materials processing must be just a means for the achievement of more basic objectives and should be evaluated in those terms. It may provide a means for increasing industrial investment to increase employment, net domestic output, overall growth, or export earnings. However, there may well be trade-offs for foreign firms between investment in processing and in extraction facilities. Thus, additional foreign investment may not become available as a result of this type of policy. Domestically sponsored investment in raw-materials processing should also face the competition of alternative projects that may be more effective in achieving overall goals.

Promotion of Political Development and National Political Goals

Political development has been identified in terms of the characteristic difficulties, crises, or problems that developing societies encounter and must resolve to become viable. These are the establishment of national identity and recognition of the political legitimacy of the governmental system, the increased participation of the citizenry in the political process, the extension of government influence more broadly into various aspects of life, and the distribution of both material and associational values. These problems must be surmounted by every political system if it is to survive, and the ways in which these problems are resolved both shape and are shaped by the capabilities of governments.[10] The rapid changes in the means used to deal with these problems give a particularly urgent quality to political development.

That political problems must interact extensively with technological decisions is apparent from the inclusive nature of the issues. Therefore, it might be possible to define appropriate technologies according to how much they help in the management of these problems. However, the characteristics of technology have not been systematically associated with political development to any great extent. At this point it is difficult to do more than use the political development problems for rationalizing technological decisions that are otherwise difficult to justify. Thus steel plants, even when not warranted on a narrowly defined economic criterion, may be an appropriate technology on the ground that they contribute to the creation of national identity. Such "national monuments" can serve an important symbolic function and, as such, may be a major political tool in assisting national integration. Public-communications methods and transportation systems not necessary to meet the economic requirements of development may be justified in terms of the capability they create for the government to spread its influence throughout the country. As noted above, the technologies associated with relatively large-scale projects may be both reflections and instruments of an intention to facilitate a "modern" centralized political development rather than a decentralized village and rural society.

Another criticism of some technologies is that they are foreign and thus maintain or increase national dependency upon external capabilities.[12] Although dependency has received a good deal of attention, especially in Latin America, it remains an elusive characteristic. It is often defined tautologically or in terms of such general constraints on national autonomy that its significance cannot be tested. Use of imported technology does mean dependence on foreign technology in a simple sense. But the social and political consequences are not necessarily negative in the sense attributed to them by the dependency arguments. Nonetheless, it is not surprising and must be recognized that, with technology as with foreign trade, partial if not complete autarky can be an important political goal.

Improvement in the "Quality of Life"

Although all the above economic and political criteria for appropriate technologies which are standards based on technology's contribution to specific aspects of economic and political development, another approach claims to base its standards on a different valuation of the major features of modern societies. This alternative view places a high value on small-scale economic activities, on rural or village life, on local self-sufficiency, on the maintenance of a "natural" ecology, and on employment and equalization of the distribution of income. These themes are, perhaps, most clearly identified with Mahatma Gandhi. However, most of the elements have a wide range of advocates, historical as well as contemporary. Among the most well-known of the latter is E. F. Schumacher, who has popularized the notion of an *intermediate technology:*

It is vastly superior to the primitive technology of bygone ages but at the same time much simpler, cheaper, and freer than the super technology of the rich. . . .[13]

The intermediate technology would also fit much more smoothly into the relatively unsophisticated environment in which it is to be utilized. The equipment would be fairly simple and therefore understandable, suitable for maintenance and repair on the spot . . . far less dependent on raw materials of great purity or exact specifications and much more adaptable to market fluctuations than highly sophisticated equipment. Men are more easily trained; supervision, control and organization are simpler and there is far less vulnerability to unforeseen difficulties.[14]

These specifications are less criteria for choice among existing technologies than aspirations for a new set of technologies. The hoped-for intermediate technologies would not sacrifice anything in terms of output and income and yet would employ more labor, use less of other resources, and so on. It is far from clear that such technologies can ever be created for a broad range of production activities. This approach to technological choice does emphasize the many disadvantages of small farmers and artisans relative to larger producers who have greater economic and political power. The search for intermediate technologies has been justified in part by the desire to reduce these disadvantages. The technological remedy proposed would essentially permit a withdrawal or at least provide protection from market interactions by permitting a greater degree of self-sufficiency. However, the existence of market imperfections and political disadvantages does not necessarily imply that technologies that compensate for the imperfections should be sought and adopted. For example, a criterion proposed for village-level technologies is that the cost of the required capital equipment should be low enough for a single farmer or artisan to afford it. This criterion is justified by the customary shortage or unavailability of loans necessary for larger-scale technologies. In effect, a financial market imperfection—the unavailability of loans to small farmers at market interest rates—is used as the rationale for a particular type of technology. But that implies adjusting to, and preserving the effects of, the financial market imperfection. It would be better economic and social policy to attempt to eliminate the market imperfections themselves instead of trying to create techniques that preserve the effects of such imperfections.

Although the Intermediate Technology Group and other organizations claim some success in finding technologies that satisfy their requirements, their examples appear to be clustered in a few agricultural activities and small manufacturing activities outside the major production processes of both sectors. In the absence of cost–benefit analyses for new intermediate technologies, it must be concluded that for the major processes in mineral transformation, power generation, and the metal and electrical industries—i.e., much of the stuff of modern life—village-level intermediate technologies remain unfulfilled goals. Perhaps a Gandhi or Schumacher would respond, first, that insufficient effort has been put into the creation of intermediate technologies to judge the potential. Yet it is not correct to argue that there have been no incentives to create such technologies. Moreover, it should not be expected that modern science is necessarily capable of doing anything, even developing an intermediate technology for producing stainless steel. A second reaction would be that man should learn to live without so much stainless steel and the other commodities of modern life that are so demanding of resources and that, to a considerable degree, impel production into large-scale, low-labor intensity, agglomerated industrial complexes. But this is a matter of choice on which the overwhelming evidence is a preference for modernity.

Thus, the Gandhian–Schumacher prescription for appropriate technology is, at this point, mainly a desire for a life-style focused around villages and a relatively simple, rural existence compared to the mass production–consumption, urbanized life-style more typical of the relatively developed countries. There is no criterion for economic choice involved because the intermediate technologists promise more of everything except what they consider undesirable in life. The arguments are not in themselves conclusive, and, if research does generate intermediate technology alternatives, choice will finally be one of individual taste. About matters of taste, there is, finally, no disputing.

These two critical aspects of the Gandhian–Schumacher prescriptions for appropriate technology—(1) that it is at present a research agenda rather than a practical set of alternatives, and (2) that if the research were successful, it would be a prescription for a different life-style—are typically obscured in debate about the characteristics and implications of intermediate technologies that do not now exist. Nonetheless, there can be little objection to the consideration of adding the intermediate technologies to the technical-research agenda for developing countries. If intermediate technologies having the characteristics described above can be found, they would improve the economic conditions of life for the majority of persons in most of the less-developed countries who are still living in rural villages. However, it is not certain that everything called an intermediate technology would have that effect. The economic outcome depends on the relative effectiveness with which such technologies would use resources compared to other techniques. Beyond that, their acceptance and use depend on other relevant criteria and the preferences for the life-styles they make possible. But there is really not much point in debating at length the merits of intermediate technologies that do not now exist and whose contrivance remains to be demonstrated.

Complementarity and Competitiveness of Alternative Criteria for Appropriate Technologies

To decide what is an appropriate technology, it is necessary to have a criterion. Since particular technologies are not ends in themselves but means for achieving economic, political, and other social goals, the criteria for appropriateness must be found in the goals of development. Yet there are many different development goals, and of these some are alternative and competitive rather than complementary. For example, maximization of output and income in any particular year may not produce the fastest rate of overall growth over a longer period. Increasing employment and reducing unemployment may or may not be compatible with either output and income maximization or maximizing the overall growth rate. Providing employment may lessen or increase inequality in the distribution of wealth and income, depending on the sectors in which employment is created. Employment creation in the short run may well compete with employment creation over a longer period if the former results in the sacrifice of investment that would create employment opportunities in the future.

Extending the regional distribution of development may also be consistent with all the other goals but, on the other hand, may be a competitive objective, depending on the reasons for the regional inequalities and the means used to reduce them. For example, if technologies chosen on other grounds create balance of payments problems, then certainly sacrifice in some other dimension will be necessary to help redress the balance of payments.

With respect to the relations between political development and other goals achievable in part by appropriate technological decisions, few generalizations appear warranted. Political criteria that emphasize the political symbolism and penetration or social mobilization values of projects may conflict with economic criteria. Certain types of economic organization and technologies may contribute more specifically to centralized political development goals than others. But, for the most part, the relations remain obscure, even though the potential for the relationships is clear.

Finally, intermediate technologies stressing small-scale, relatively self-sufficient production and village, if not rural, life are competitive with social and economic goals associated with urbanization. However, the reality of the choice cannot now be assessed. There are undoubtedly economically efficient technologies in some sectors that are intermediate in the sense that they employ capital and labor in proportions different from those used by modern large-scale industry. However, it is by no means clear that even these satisfy the Gandhian goals. Of those specific technologies proposed for these latter purposes, few, if any, have been carefully assessed with respect to either their narrowly economic or more broadly social characteristics.

When goals and criteria are consistent and complementary, technological decisions are relatively straightforward. But what can be done when the goals of development and criteria for decisions about appropriate technology are inconsistent and competing? That is a complexity of reality that cannot be escaped and for which there is no analytical resolution. For example, suppose it is conclusively demonstrated that some techniques are more labor intensive, yield as high a return on capital as alternative methods, but generate lower savings. This would imply a lesser contribution to further investment and growth. The different goals are competing and must, in effect, be valued against each other. An essentially political decision is required in determining the extent to which one goal will be pursued at the sacrifice of others.

Moreover, to determine which criteria of appropriateness will be observed or the relative weights each will have, it is not sufficient to make an initial decision. When there is competition among the criteria, the competition will continue through the lifetime of the project. In particular, cost minimization is the only criterion that is surely consistent with the survival of the private enterprise that employs it. That is true as well of public enterprise required to meet a market test. The technical decisions of a government-operated railroad system, for example, can be made primarily to provide employment, rather than to minimize costs. But suppose also that the government system has to compete with a private road–trucking system. Since the government railroads do not minimize costs, they would lose business to private trucks, and only recurring government subsidies would compensate for the employment-intensive but high-cost railroad technology.

Enterprises that must directly or indirectly meet market tests of performance and do not use cost-minimizing technology will require continuing government intervention to guarantee their survival. Thus, before any criterion of appropriateness other than cost minimization is actually applied, the government's ability and willingness to provide the continuing support required by the technology should be appraised. Choice is easy; survival is more difficult.

REFERENCES AND NOTES

1. For an example of planning to achieve future targets see *Third Five Year Plan*, Government of India Planning Commission.
2. See P. Diamond and J. Mirrlees (1971).
3. See, for example, R. Eckaus and K. Parikh (1968), Chap. 1, and P. Dasgupta *et al.* (1972).
4. This terminology is due to the elucidating discussion of A. Sen (1975), Chap. 1.
5. E. Schumacher (1973), *passim*.
6. See Chenery *et al.* (1975), Chap. 1.
7. See G. Ranis (1974) and H. Hughes (1974).
8. For a review of the debate that favors export promotion see Little *et al.* (1970).
9. For a review of studies of this question see United Nations Conference on Trade and Development (1975), pp. 57–59.
10. For a discussion of these characteristic political development crises, see L. Binder (1971).
11. See N. Choucri (1976).
12. See, for example, O. Sunkel (1973).
13. E. Schumacher (1973), p. 145.
14. *Ibid.*, p. 176.

4 Technological Opportunities and Transfer of Technical Information

Introduction

In attempting to make and implement optimal decisions with respect to appropriate technologies, the acquisition and transfer of information about technological alternatives is, in itself, one of the foremost difficulties. While in many cases there are obviously a number of alternative sources and channels, in other cases there is only one or a few. And in most cases, information about the costs and returns of the alternatives is seldom easily ascertainable. This chapter will survey the state of information with respect to technological alternatives and the methods of transferring technological knowledge, to provide background for the subsequent examination of the various processes of, and influences on, decision making.

Range of Technological Alternatives from which Choices Can Be Made

If only one production method for any particular type and quantity of output were physically efficient, then the problems of technological choice would be greatly simplified, though not eliminated. With no alternatives, there would be no need to consider criteria for choice unless physical efficiency were willingly sacrificed. Even so, by considering diverse products, each requiring a single but different set of inputs, it might still be possible to achieve a range of choice among alternative resource–input combinations. Assessment of this last possibility requires information not only about technology but also about demand patterns. Choosing an output mix to get an appropriate input mix is pointless if the outputs cannot be matched to demands.

It has been hypothesized that in many of the important economic sectors there is, in fact, only one, or at most a few, alternative methods for carrying out the central production processes and that all methods require inputs in much the same proportions.[1] Such a limited range of alternatives, it has been suggested, does not permit the physically efficient employment of all the labor that is available in some developing countries because of their limited amount of capital and other resources. If this strong hypothesis about technology and labor and other resource availabilities were correct, then some unemployed labor simply could not be usefully put to work to produce additional output. A variation of this single-technique hypothesis applies not to the entire economy but to some particular sectors, usually utilities and manufacturing, or at least the heavy industry part of it. The single-technology view appears most prevalent among engineers who think in terms of a best-practice technique that is virtually independent of wages and prices of other resources.

Even if only a single efficient combination of labor and other resources could be used to produce any output, it would still be possible to employ labor or other resources more intensively by not using them in a physically efficient way. Where one man could do the job, two or more could be employed. Labor and other resources can be used in many inefficient ways, and in certain conditions inefficient use of labor might be desirable if that were the only way to achieve some other goal such as, for example, income redistribution. Such a conscious decision should, however, be distinguished from the many instances in which inefficiency is not recognized and not desired.

The more conventional view of technology, within the economics profession at least, is that many different combinations of inputs can be used to produce more or less the same output. This production hypothesis implies that one input can generally be substituted for another over a wide range of input proportions to achieve any particular level of output. If this view of technology is correct, then the technology, though still having great economic influence, does not so completely determine the quantities of resources used in production as in the single-technique hypothesis. Moreover, if the conventional hypothesis is correct, a different explanation of unemployment in developing countries is implied. Aside from the problems of effective demand, unemployment must be the result of rigidities in the wage-payment mechanisms, such as minimum-wage laws imposed by government, union wage agreements, or by some other departures from the competitive markets. Otherwise, labor could always be substituted productively for other resources.

Clearly, to understand and to make policy in developing countries, it is important to know the range of substitution possibilities and the relative trade-offs. As a result, substantial attention has been devoted to empirical investigations of the issues. These investigations are unusually difficult and often costly. Detailed information on the use of resources exists only at the plant level, and even there it is often not conveniently organized. Manufacturing census data are usually too aggregated and incomplete to reveal the type of technological information required, so that special case studies and surveys are necessary. However, detailed production data is often difficult to obtain because of proprietary interests. Therefore, it becomes necessary to resort to indirect and less-penetrating methods of investigation. As a result, the studies that have been undertaken are, unfortunately, inconclusive, although information is slowly accumulating about particular sectors and production processes.

One kind of empirical investigation of the opportunities for technological choice has relied on statistical information from censuses or surveys about the resources actually used in production. The hope has been that inferences about technology could be drawn from information about resources used. Yet the information often includes only the amount of resources used, employment and wages paid to labor, the profit share, and, perhaps, estimates of capital stock. Typically, these sources also contain only highly aggregated information on products that are not strictly identical. The data may cover different periods of time during which significant technological improvements may have been implemented, although at uneven rates. As a result, much ingenuity and many restrictive assumptions have been necessary to produce any measures of the potentiality for substituting inputs for each other, typically capital for labor or vice versa. Despite the ingenuity and because of the inadequacies of the data, the results are inconclusive. Some studies seem to show only a restricted range of alternatives, while other studies indicate a wide range of easy substitution possibilities between capital and labor as potentially the most important trade-off. Furthermore, none of the findings appears to be "robust"; apparently minor changes in estimating procedures will lead to substantial differences in the results.[3]

Another type of attempt to establish the range of available technological choices has involved detailed case studies of particular types of production processes. In some cases, the study has been made at a highly disaggregated level and has proceeded by isolating subprocesses within the overall production of a particular commodity. In other cases, the study has not tried to penetrate below the level of what goes into and comes out of the factory or farm. One especially assiduous technique requires visiting and surveying individual plants. The case-study methods are relatively slow and expensive, though the results are often not only insightful for the the specific methods and products but also suggestive with respect to a wider range of technologies.[4] By their very nature, however, the case studies produce piecemeal and inconclusive evidence. In some cases, the advantages of labor-intensive techniques appear to have been demonstrated; in other studies, the most labor-intensive techniques do not appear to be the most appropriate for minimizing the cost of production. An interesting suggestion from some investigations is that, even if there are only a few efficient technological alternatives in the central manufacturing processes, significant substitution possibilities exist in the "peripheral" materials handling and transport functions.[5]

A major problem in interpreting the typical case study of particular establishments and industries is the difficulty of determining from the available evidence whether the observed alternative input ratios are all efficient in the least-cost sense. As noted above, apparent differences in input proportions for the same type and quantity of output can be generated simply through inefficiency. Presumably, few establishments set out to use resources inefficiently, but many of them might be doing so, as suggested by the previously noted wide and prevalent differences between actual and best-practice techniques.

While it is important to know whether the methods observed through case studies reflect efficient resource use, such information can only be obtained by comparisons beyond the scope of such studies. The problem typically is avoided by the implicit or explicit assumption that the production establishments studied are participants in perfectly competitive product and resource markets. If this were the case, then the observed labor and other intensity resources could be considered to have "passed the market test" and, therefore, to be efficient. However, the assumption is seldom if ever tested explicitly, so the conclusions of the case studies about efficient alternatives must often be accepted only with reservation.

The economies of scale characterizing many industrial processes are a boon because of the relatively low cost they make possible. But they also create problems, because developing countries are typically less able than industrialized countries to take advantage of economies of scale. The lower costs of output often associated with larger plants can be achieved only if the plants are operating at near capacity. This means that many developing countries with small markets due to their relatively small populations and/or low per capita incomes cannot achieve the lowest existing production costs if plants are built to serve the domestic markets alone. If national self-sufficiency is set as a policy goal nonetheless, economies of scale mean that developing countries with relatively small plants will bear higher real costs of production.

International trade provides a potential method of using the capacity of large plants through exploitation of foreign markets. However, costs and time delays in penetrating foreign markets make it difficult for the newly industrializing countries to displace the production of advanced countries.

If international financial markets were more perfect, the developing countries would not be at a disadvantage just because of the large amounts of financing needed to create firms that can achieve the significant economies of scale. However, financial markets are not perfect, and large loans to finance large plants in small countries may be considered particularly risky. For these reasons, large size and the consequent large amounts of funds required can, in itself, be a barrier to low-cost production by developing countries.

The evidence is not conclusive for one feature often attributed to production processes characterized by economies of scale. That feature is the argument that such processes are also relatively capital intensive rather than labor intensive and, on this ground, unsuited for developing countries. What typically captures the eye and generates such observations about large plants is the scale of plant and equipment, not the relative amounts of factors used in efficient production, which is the relevant issue. It may be suggested that output levels rivaling those of large-scale plants could be achieved by replicating more obviously labor-intensive methods. However, observations of the naked eye will not guarantee that replicated labor-intensive methods are as efficient as large-scale methods. The latter may conceivably be the most labor intensive of the efficient techniques that are available. Unfortunately, the lack of detailed technical knowledge again impedes a more definitive statement.

In studies of the potential range of input substitution, special attention has been given to the agricultural sector. Here there are dramatic examples of diversity in the relative intensity with which various inputs can be used: in labor and machinery use, in fertilizer, in irrigation water, in land leveling and improvements, and so on. The existence of an important range of technological alternatives would, in fact, appear obvious. The question of whether the observed alternatives all use resources efficiently is typically answered by pointing to evidence suggesting that farmers respond to the incentives of output prices, even in traditional peasant agriculture, and thus that the markets in which they operate are competitive.[6] This evidence, however, is not definitive. Responsiveness to price is a necessary, but far from sufficient, condition for the existence of reasonably competitive markets, and there is other evidence of less than the full participation in markets required for economic efficiency.[7]

Another approach to studying technologies is to carefully build up the necessary information from engineering sources. In a variation of this approach, engineering and economic information is accumulated to design "synthetic" plants. The research methods are potentially among the most powerful, since they require fewer assumptions about the physical efficiency of observed technologies. However, the research approach generates contrived production methods that may not have faced a market test. The evidence generated by this "engineering–economic" approach is again equivocal. A wide range of efficient substitution possibilities appears to exist in some cases and not in other cases.[8]

This survey indicates that the existing information on the range of economically efficient technologies is not conclusive. But some impressions do emerge. For example, in the light-manufacturing and agricultural sectors in which most effort has been concentrated, the results seem typically to suggest that, by the criterion of employment, there is clearly a range of alternative technologies in many sectors. Most of the case studies of particular industries demonstrate this, as does much, though not all, of the evidence based on statistical investigations of census or other data collections. The latter type of evidence may, to some extent, reflect differences in labor absorption possibilities in the manufacture of somewhat different products or similar products of somewhat different qualities. As noted, it has generally been difficult in the statistical investigations to ensure that the information used pertains to only a precisely identified product.

Yet there should be no surprise that the empirical tests show wide ranges of employment possibilities. The studies of the substantial differences between actual and best-practice techniques have often focused on the differences in labor productivity among firms. That implies differences in potential labor absorption, but not necessarily that the alternatives are either physically or economically efficient.

Virtually all the empirical studies of technological alternatives have focused on inputs and their relationship to output. To the extent that the other criteria of technological appropriateness discussed in Chapter 4 are different from efficiency or employment standards, there is little evidence about the existence of a range of alternatives.[9] Examples in the handbooks on cost–benefit analysis that use the criteria of economic growth or the maximization of the availability of consumption goods indicate that by these standards there are differences among projects.[10] But the examples are not usually directed toward the analysis of different methods of producing the same product, nor are they actually (nor intended to be) surveys of the range of technological alternatives.

A great number of "intermediate" village-level technologies have been proposed; however, for the most part, these proposals consist of a specific design intended to demonstrate technical feasibility. That is, the proposals characterize the technologies in terms of required inputs and related outputs. Only a few of the proposed intermediate technologies include an economic cost–benefit analysis, and virtually none have been analyzed according to other criteria of appropriateness.[11]

For the goal of income distribution there appears to be no systematically accumulated and analyzed evidence on technology other than those effects that operate via employment. There may be a number of other relations between technology and income distribution such as the potentially important differential skill requirements of different technologies. There may also be systematic differences in the average age of the labor force using new technologies as compared to the labor force using older technologies, and that would have distributional effects.[12] However, no systematic information of this sort is now available.

As noted previously, correcting the balance of payments is typically not seen as a goal distinct from the goals of output maximization or cost minimization for particular output levels or overall growth. Nonetheless, it has been argued that different technologies have been employed in the pursuit of policies to promote exports as compared to those designed to substitute for imports. This, it is claimed, is the result of differences in effective incentives.

The fact that technology influences political development has also been recognized. It is easy to fall into the belief that modern technologies of communications and control make it easier for governing elites to establish their influence at all levels of their societies. Yet, such influence has been achieved even in societies with only relatively primitive technologies,[13] but these have been traditional rather than modernizing societies. It has also been pointed out that modern technologies increase the effectiveness of contention by dissenting groups within a society.[14] The possible relations between the degree of emphasis on large-scale industry and its associated technologies and modern political development have been discerned but not detailed. So, again, few generalizations about the relations of particular technologies to political development now appear to be warranted.

Finally, choosing technologies to improve the quality of life raises not only the question of different tastes in life-styles but also questions about the objective relationships between particular technologies and specific aspects of the quality of life. It is the latter set of issues that will be taken up here. It is clear that for many products smaller-scale, less-polluting and more labor-intensive technologies can be used instead of those now widely employed. What is almost totally lacking in the recommendation of such technologies is an overall appreciation of the implications of their widespread adoption. For example, animal-drawn conveyances *are* smaller scale, more labor intensive, do not emit sulfur or nitrite pollutants, do make use of "renewable" resources, and are more "natural" than vehicles run by motors dependent on fossil fuels as energy sources. Yet the large-scale use of horse-drawn vehicles was a major source of urban pollution before the advent of the automobile, and large concentrations of farm animals are a continuing water-supply-pollution problem in some rural areas. Similarly, thermoelectric central generating stations can be obvious and concentrated sources of pollution. Yet, they do burn coal and oil more efficiently than the small boilers that would be used in small-scale and more dispersed generating stations or steam-producing units.

With respect to the potential for technological substitution, the advocates of village-level intermediate technologies promise that major research efforts will increase the range of techniques that can be used with "reasonable" efficiency to improve the quality of life. That remains to be demonstrated.

As a research agenda, the village-level intermediate-technology approach is not substantially different from the motivation of much of the work now being done in many places. Many of the developing areas have their own scientific and engineering research institutes engaged in efforts to find technologies better suited to the local resources, and there are a few international institutions of this type. They are attempting to extend the range of known technologies into unknown territory, to discover more appropriate methods. The Agricultural Engineering Department of the International Rice Research Institute provides an outstanding example of organized efforts to develop appropriate mechanical technologies for rice cultivation in Asia. The standard of appropriateness in this and similar institutions appears to be lower cost or larger output for the same cost. Only seldom is another standard used to evaluate the new methods generated in the various research organizations. That is primarily because, for new technologies as well as old, it is difficult to establish clear relationships between a particular technology and the other standards of appropriateness that have been suggested.

Sources and Costs

Information in developing countries about technologies may have its source in new research and development, transfer of knowledge from other areas, and/or adaptation from existing methods.

Information about existing technologies can be transferred from a number of sources through a variety of channels.[15] Technical information can be obtained via sources that are virtually "free" in the sense that little, if any, cost is involved in obtaining access. This is essentially the case for the information found in technical books and journals. But using that information requires a "processer," an engineer or scientist who understands the literature. Since trained scientists and engineers are relatively scarce in the less-developed countries, this source is less readily accessible than in the industrialized countries. However, once someone with the appropriate training is available, the information contained on the printed page is available with only small charges up to the "processing capacity" of the technical interpreter. There are other sources of free information in addition to the printed page. For example, the knowledge that a particular type of production is feasible in a developing country is free and cannot be hidden if there is any production at all. While obvious, it can be quite important. Educational institutions often provide information which, if not free, is of relatively low cost. Casual conversations and structured technical meetings will similarly be occasions for virtually free exchange of knowledge.

How much technological knowledge is freely available or at low cost, and how significant is it? How much of that is general background information, and how much is "hard" technical data? There is no basis for arriving at an assessment.

Domestic research and development is carried out in both private and public institutions in developing countries, but typically on a small scale except for agricultural research. The accomplishments of the industrial research and development organizations, while apparently numerous, have seldom been evaluated for their significance.[16]

Another repository of a great deal of technological knowledge is people. The scientists and engineers who learn formally in the schools and universities are perhaps the most obvious examples. But the foremen and craftsmen and even the unskilled laborers "carry" a good deal of detailed information, ranging from specifically technical methods to general knowledge of work organization. It may be acquired "on the job" or in "vocational" courses. Some is job specific, but some is to a degree transferable. Formal education is acquired in many of the developing countries by sending students abroad, particularly at the university level. Internal mobility is also a way of spreading technological knowledge within a country. However, in developing countries the kind of technological knowledge required by industrial foremen and lower-level workers is not easily acquired. Such workers have few opportunities to travel and gain experience abroad and only limited opportunities to learn from experience in their own countries because of the relatively small amount of industrial employment.

In many cases, the necessary knowledge for a production process is protected by foreign patents that are protected by international patent agreements. In this case, to use the patented information the enterprise in the developing country must pay royalties or license fees. These fees are determined in bargaining processes in which the representatives of the developing country are likely to be at a disadvantage in facing the monopoly power of technology sellers. The results are often agreements that restrict how the technologies are used and/or where the output is sold.

Quite apart from patented technology, some information is available only with foreign participation of some kind, often the direct participation of a multinational corporation. These organizations, which have attracted a great deal of attention in recent years, are undoubtedly carriers of much information, some of it protected by patents, some of it unpatented know-how embodying the technical and managerial expertise accumulated in the organization. Given the special international operating capabilities of the multinational organizations, they have been able to exercise important types of influence in many of the developing countries. Assessment of their decision-making powers and their influence will be discussed in the next section.

As noted, much of modern technology is embodied in particular machines. The machines themselves may not even be patented if they operate on well-known principles. But to have the production capacities these machines can provide, it is necessary to have the machines, and those may be available in turn only from a restricted range of producers, perhaps all of them abroad. Even when the general principles on which the machines operate are well-known, there are likely to be many specific designs to resolve specific problems, and these designs may be best obtained from experienced producers.

That developing improved methods by adapting technology is widespread and important can hardly be doubted both from anecdotal evidence and systematic case studies. The classic "learning-by-doing" example of the Horndal Iron Works in Sweden is frequently cited.[17] But the improvisation of the Punjabi mechanics and engineers is equally classic.[18] It might appear reasonable that the adaptation processes in production establishments would be especially subject to the incentives of local wages, capital, and material costs. By this reasoning, the adaptation process should generate methods especially likely to "minimize" cost by more intensively using relatively abundant resources.[19] There is evidence, again anecdotal, that this happens.[20] Yet it need not follow that adaptation always moves toward increased use of the relatively abundant factors such as labor in developing countries.[21] Engineers should be interested in cost saving whenever and wherever possible. There is little evidence that adaptation in developing countries is more likely when it results in the use of more labor and less capital, rather than the reverse, or when the result is the use of less of both.

This brief review of the alternative sources of technological information suggests the difficulties of arriving at an overall assessment of the importance of any one channel, though evaluations have been undertaken with respect to specific technologies.[22] There are no statistics of knowledge flow; there are no measures of the relative importance of production based on indigenously generated information compared to foreign information. There is no way of measuring the significance of knowledge obtained from books and journals, or knowledge gained through the education abroad of scientists and engineers, or knowledge embodied in purchased equipment or in patents or other special production techniques licensed from abroad, or knowledge brought to the developing countries by multinational corporations. Thus, we are left with only general impressions of the importance of foreign technological knowledge. For example, agriculture, which is still by far the largest sector in most of the developing countries, is still the sector least penetrated by multinational corporations in most of these countries and, probably, the sector that generates the smallest amount of patent fees or royalty payments. Only a limited amount of farm equipment is used in most developing countries, and that equipment is usually not technologically sophisticated enough to command important royalty fees. That may, however, not be true for the fertilizer and insecticide plants, which will become increasingly important in supplying the agricultural sectors of the developing countries. And biological as well as mechanical innovations have been of enormous importance in many of the less-developed countries.

With some exceptions, such as transportation and power generation, the service sectors are not major importers of foreign technologies. It is in manufacturing and mining sectors where foreign technologies embodied in equipment or paid for by patent fees and royalties are likely to be the most important. Since these are also likely to be among the fastest-growing sectors in the course of development, their expenditures for knowledge will be a growing share of total costs of acquiring knowledge in the developing countries.

Unfortunately, there is no organized evidence and little conventional wisdom on either the potential for substituting different sources of knowledge or the costs of the various sources. While monopoly power controls some knowledge flows, the significance of such power cannot be fully assessed for any developing country.

Role of Engineering Education in Generation and Transfer of Technology

It has been argued that one of the major sources of bias in the generation and choice of technologies used in developing countries is the character of the education received by their engineers. Many of the engineers in developing areas still receive their education in the universities of the industrialized countries. Moreover, the engineering schools in the developing countries, because of the sources of the training of their faculties and the technical assistance received in their establishment, are often strongly influenced by the intellectual patterns of the industrialized countries. These conditions, it has been argued, bias engineering training in favor of the technological methods of the advanced countries.[23] It is claimed, therefore, that engineers from developing countries trained in these molds receive "inappropriate" education.

These are plausible arguments, but, as yet, there have been few penetrating studies of the issues. Presumably, engineering education in the industrialized countries focuses on the methods and approaches relevant to those countries. It would be reasonable to expect, then, that engineers in industrialized countries come to prefer those technologies that economize on the relatively scarce labor of those countries and use more capital more intensively. But it would be hard to document this expectation or its effects with any details. It is easier to establish that engineering education in industrialized countries emphasizes the sectors important for them, which include such "high-technology" areas as petrochemical processing and electronics.

The argument that the "output mix" of engineers from the universities in the industrialized countries, in terms of their specializations, does not match the requirements of the developing countries is, *prima facie*, more persuasive. Yet, again, generalizations are dangerous. The various fields must be examined separately to assess their appropriateness. Aerospace technology and design of high-speed digital computers, which attract a substantial number of engineering students in industrialized countries, are of little direct relevance to the less-developed countries. But nuclear power engineering, though among the most sophisticated of modern engineering fields, may also be the most relevant type of education for design and manning of the central power stations in some of the less-developed countries. Similarly, engineering education in petrochemical technology, though not relevant for many of the less-developed countries, is highly appropriate for others.

The limited opportunity in developing countries for engineers trained for high-technology sectors has been regarded as a major source of the "brain drains" that have occurred, and these, in turn, have been regarded as another bit of evidence of the inappropriateness of engineering education. Yet this may represent a mistaken identification of cause and effect. Students from less-developed countries who enter fields with limited applicability in their homelands may, in fact, be responding to international demands for engineers rather than to their own nations' demands. The engineering institutions of industrialized countries may be only a convenient scapegoat for the inclinations of engineers from developing countries to work where they can earn the highest incomes. Moreover, it is conceivable that the "investment in human capital" in departing specialists provides, for some developing countries, good returns in the form of emigrant remittances.

REFERENCES AND NOTES

1. See, for example, R. Eckaus (1955).
2. A. Kelley *et al.* (1972), p. 25.
3. For a review of the estimates of the elasticity of substitution see D. Morawetz (1975) and J. Gaude, "Capital-Labor Substitution Possibilities," in A. Bhalla (1975).
4. Two collections of such case studies are A. Bhalla (1975) and C. Timmer *et al.* (1975).

5. G. Ranis (1974).

6. T. Schultz (1964), Chap. 3.

7. The anthropological study most commonly cited in support of the "poor but efficient" characterization of peasant agriculture, itself describes the limited participation in resource markets. See S. Tax (1953).

8. V. Smith (1961) and W. Leontief (1953).

9. For example, even one of the most comprehensive analysts of intermediate technologies considers only employment, output, and trade aspects and neglects growth considerations. See G. Ranis (1974).

10. See P. Dasgupta *et al.* (1972).

11. A survey of more than 100 intermediate technologies that have been described in relatively readily available literature indicated that only 24 of these had any economic information associated with them, and for only a few was it indicated that any sort of cost–benefit analysis had been done. No analysis was presented for any of them of any qualities other than their output and employment characteristics.

12. See R. Nelson and V. Norman (1973).

13. See, for example, K. Wittfogel (1957), p. 54.

14. See J. La Palombara in Binder *et al.* (1971).

15. See F. Machlup (1967), Chap. 8.

16. For example, "Both the arguments and the evidence for an active policy of supporting the establishment of an industrial R & D effort in an LDC continue to be sketchy." R. Nelson (1974), p. 75.

17. K. Arrow (1962).

18. E. Staley and R. Morse (1965), p. 179.

19. C. Kennedy (1964).

20. N. Leff (1968), Chap. 3 and 4.

21. P. Samuelson (1965).

22. For example, L. Nasbeth and G. Ray (1974).

23. F. Sagasti (1973).

5 Determinants of Technological Decisions and Their Appropriateness

Introduction

The decisions about technologies actually used in the developing countries have been made within many different institutional frameworks and are subject to a variety of incentives. In some cases, though implicit objectives are being pursued, no explicit criterion of appropriateness governs the decision. In other cases, the objectives of the technological decision are quite clearly stated. To understand and help implement appropriate choices of technology, it is necessary to identify and assess the processes through which decisions are actually made. The purpose of this chapter is to present an analytical review of these processes. First, the role and consequences of private decision making by enterprise under market influences will be evaluated. Next, the impact of government policies on private enterprise and operating directly through public enterprise will be discussed. Finally, the international influences operating through private multinational corporations and through bilateral and multilateral official institutions will be assessed. In each case, the object will be to describe the criteria of technological appropriateness that are being applied and the manner in which they are implemented.

Technological Decisions by National Private Enterprise

Technological choices are determined solely by market conditions only under special and rigorous conditions. Conventional economic analysis shows that, when the enterprises are profit maximizers and participate fully in resources markets, they will tend to make technological decisions in such a way that the incremental value of the output from using more of any one input will equal the extra cost of using that input. This description of technological decisions is a rather abstract one and does not describe any particular technology. Nonetheless, it is a powerful characterization and permits another step in the analysis. Suppose that the resource markets in which enterprises hire labor and buy other inputs and the product markets in which they sell are both perfectly competitive. Then, if decisions are made as described, firms will maximize not only profits but also the total value of output the system can produce. Moreover, under the conditions cited, and if all capital resources were used, there would be no unemployment of labor. The wages of workers would be flexible and would always adjust so that it would be profitable to hire unemployed labor. Thus, if competition is perfect, and if it is always possible to use labor productively, two appropriateness criteria would be satisfied simultaneously—the value of output would be maximized and full employment would be achieved.

However, even under the extreme assumptions of maximizing behavior and perfect competition, it is not possible to conclude that the technological choices would be satisfactory in terms of any or all of the other criteria described in Chapter 3. The invisible hand guiding individual decision making to the social goals of output-value maximization and full employment will not guide even a competitive system to goals that are not expressed and fully decided through markets. Furthermore, economic growth and income-distribution goals and even the balance of payments correction goal are not usually market determined, nor are political goals or general quality-of-life objectives.

Maximizing behavior by private firms in competitive markets will not produce the socially desired growth rate for a developing country for a variety of reasons.[1] The growth rate is constrained by economic conditions, but within those constraints it is essentially a political decision, in the broadest sense, and depends on the resources that can be mobilized by the public sector as well as the private sector.

The effects of a competitive system on the personal distribution of income are not readily determined. Although there is a growing literature on the relations between market structure and income distribution, few conclusions will bear much weight at this point. For distributional objectives, as for growth objectives, it can be said that a competitive system does not guarantee the achievement of the particular distributional goals that developing countries set for themselves.[2] This general conclusion must also be true for the political development and the quality-of-life criteria of appropriateness, though such issues have seldom been considered.

The blindness of competitive systems to distributional and noneconomic goals is a well-known feature of such systems. That does not mean that they do not have implications for income distribution, regional dispersion, the quality of life, political development, and so on. But those implications have not yet been identified by a general economic or more broadly social analysis.

At the other end of the economic spectrum—with perfect monopoly—and still with profit-maximizing behavior, even less can be said, except that the output value may not be maximized for the entire economy. There would still be full employment if the other conditions cited prevailed, but workers would not, in general, be so productively employed as if there were competition. However, with respect to the other economic and noneconomic goals embodied in the technological appropriateness criteria, little if anything general can be said.

Complete monopoly may be as rare as perfect competition, especially when the potential for substitution of alternative types of goods and services is taken into account. However, it is especially likely that many markets in developing countries suffer from some degree of monopoly power that controls prices by controlling levels of output and sales. Important economies of large-scale production in many types of manufacturing industry lead to lower costs as the level of output increases. These economies may stem from the basic physical transformation processes as well as from the time, labor, and equipment savings associated with running continuous production lines or large-size batches. While such cost savings do not necessarily increase indefinitely with the level of output, the minimum unit cost often occurs at scales beyond the size of the markets typical of many of the developing countries. Thus, it is difficult for several firms to coexist if they compete actively. The markets for many manufactured goods in the developing countries, in turn, tend often to be small, not only because of low incomes and small populations but also because of relatively high transport costs that divide the country into somewhat separate consumption units.

In the spectrum between competition and monopoly, many kinds of markets are possible. Thus, few if any other generalizations can be made with respect to output, employment, or the other criteria of appropriateness. However, there is no reason to expect any of these goals to be achieved by economic systems characterized by substantial market imperfections due to important elements of monopoly power.

Yet the market imperfections associated with the existence of monopolistic elements are not the only ones that characterize developing countries. Market fragmentation or partitioning is a common feature of many of their markets. Market fragmentation occurs when particular groups of actual or potential buyers and sellers are segmented into self-contained markets and do not interact with other groups. An extreme example is when a producer simply does not participate in markets as a buyer of productive resources or as a seller of his products. Complete self-sufficiency by peasant farmers would be such a case. The restrictions on sale or rent of land and hiring of labor that characterize traditional agricultural practices in many areas are examples of less-complete self-sufficiency.

The fragmentation or partitioning of markets has a variety of sources. In financial markets it is, in part, due to the official restrictions placed on chartered financial institutions that limit the range of their financial activities. In agriculture and other sectors where traditional patterns of behavior may be especially important, these behavior patterns will restrict the kinds of transactions that are socially acceptable. While such patterns can and do change under the pressures of development, they are also quite persistent, and it cannot be assumed that change will necessarily be swift.

Market fragmentation limits the effectiveness of economic forces in achieving full and efficient use of labor and other resources. Unemployed and low-productivity use of resources in one sector may exist simultaneously with nearly full and efficient use of similar resources in other sectors.

These observations on the pervasiveness and effects of market imperfections are significant partly because two quite different kinds of recommendations have been based upon them. It was noted earlier that widespread monopoly power, which especially impinges on small and rural producers, has been cited as the justification for the search for small-scale technologies. But these technologies, if achieved—and there is as yet little evidence that they can be—would permit greater self-sufficiency, so that essentially they represent withdrawals from market participation. But this would substitute one kind of imperfection for another. The other argument, which has come with force and repetition from authoritative sources, is to allow the market to solve problems of unemployment. According to proponents of this recommendation, individual enterprises choose the appropriate technology in response to prices determined in efficient markets.[3] The recommendation is based largely on the apparent success of several countries in avoiding a major unemployment problem in the course of development.

Korea, Taiwan, Hong Kong, and Singapore are the usual cases cited. The argument starts with the claim that a significant range of alternative, efficient, labor-intensive techniques exists. However, other than the studies of substitution possibilities discussed above and finally viewed skeptically, the evidence cited is indirect: relatively intensive use of labor is observed. From this observation, the proponents conclude that the standards of efficiency are satisfied. But this argument is based on the assumption that these markets are reasonably free from noncompetitive elements and are not fragmented and that managers are profit maximizers.

While we are not attempting here to evaluate the conclusions for the countries studied, their limited validity is suggested by the contradictory observations of the narrow scope for capital–labor substitution and of market fragmentation. Moreover, other studies have generated the paradox that, even when relatively labor-intensive technological choices exist and are more profitable than capital-intensive methods, the labor-intensive methods are not always chosen. In a study of technological choices in Indonesian manufacturing, this choice has been attributed to an "engineering bias" for modern, capital-intensive techniques. Interviews supplemented by some quantitative data indicated that decisions made on technology were not dictated by cost-minimizing considerations alone but were heavily influenced by and, in some cases, finally decided on technical grounds that required the sacrifice of profits. The investigator concluded that:

While part of the drive toward capital-intensive technology may be explained by the desire of the oligopolist to insure against risk and uncertainty, a large part seems also to be a response to some objectives of the engineering man. These are:

1. Reducing operational problems to those of managing machines rather than people.
2. Producing the highest quality possible.
3. Using sophisticated machinery that is attractive to the engineer's aesthetics.[4]

Yet, such a skeptical study may be no more generalizable than the results of the studies on which the "let the market work" recommendations are based.

A review of the available evidence led one observer to conclude:

Unfortunately, we cannot sort out to what extent the use of capital-intensive technologies can be attributed to (a) the unavailability of labor intensive alternatives . . . , (b) inappropriate conditions of choice (information, prices, etc.) . . . , and (c) irrational preferences for sophisticated methods of production. . . .[5]

That unemployment is not a major problem in particular countries does not in itself necessarily imply the effective working of good labor and other resource and product markets. Yet skepticism about the power of unimpeded markets to resolve unemployment problems in the developing countries does not imply that markets do not influence the use of resources and that it is not necessary to be concerned about market distortions, whatever their source. It is important that market as well as other incentives point in the right directions, if markets are to be used to mediate economic decisions. Otherwise it is unlikely that the development goals will be achieved. However, correct market incentives may not be a sufficient condition for that achievement.

Influence of Government Policies on Choice of Appropriate Technologies

Government intervention in the economies of the less-developed countries ranges from the direct ownership of some production enterprises and direct controls over private enterprise to the use of a wide variety of taxes and subsidies to accomplish overall objectives. These policy tools are used to achieve overall as well as distributional and welfare goals, in some cases by directly influencing the choice of appropriate technologies. But in many instances, the effects on such technological choices are unintended by-products of government policies.

From their observed actions, it appears that governments pursue different goals at different times and, by the same or diverse means, at the same time. Some government projects have been used to provide employment; others are carried out and operate to maximize output or to minimize cost. In community-development projects, governments have attempted to directly shape the quality of life of the participants. While it would be helpful to know the relative weight that government decisions give to the different appropriateness criteria in particular countries, that information is seldom if ever explicit, and no studies have yet elicited it.

Finally, an important point from Chapter 3 should be recalled. Government, by its direct action or control over private investment, can impose on enterprises any criterion of technological appropriateness it desires. But only if the goal is to maximize net revenues will the enterprise have reasonable hope of viable independence from the government budget. If any other criteria of appropriateness are pursued, it is likely, and in some cases certain, that the enterprise's revenues will not cover its costs. This is simply because pursuit of other goals means that the enterprise is not trying its best to cover costs. And, if the enterprise cannot cover its costs from its revenues, it must be subsidized from the government budget if it is to continue to operate.

The direct government controls over new investment that are common in the developing countries are designed to achieve sectoral allocations that are more effective in contributing to the desired rate and pattern of development than if the investment decisions were determined by private enterprise and market influences. By controlling the distribution of investment between the consumption and capital-goods sectors, these direct controls also help force a higher rate of saving than would otherwise be achieved. In some cases, direct control is exercised over the use of specific technologies as, for example, in determining the scale of production or the type of fuel or raw materials used. Except in some public enterprises, direct controls over output are rarely used in the developing countries, though agreement on specified output and export targets may be a condition for the granting of investment licenses.

Price controls over both inputs and outputs are common. For example, minimum wage laws are intended to offset local monopoly power in hiring and to assure minimum incomes. Maximum prices on certain foodstuffs have similar objectives in the subsidization of consumption. Regulations that set maximum interest rates are intended to encourage investment and, when they are set differentially for loans in different sectors, to direct the allocation of investment.

In addition to these direct quantity and price controls, the developing countries have much the same arsenal of taxes and subsidies as those used in the industrialized countries. These taxes and subsidies extend to foreign trade in the form of tariffs and export subsidies. In the foreign-trade sector, various types of quota restrictions over imports are also used. Although revenue collection is the major objective of many of the taxes and subsidies, some are used to influence patterns of output or use of inputs. The effects of all these fiscal and quantitative instruments depend not only on the structure of each of the developing economies but also on their abilities to assess and collect the taxes and to control the payment of the subsidies.

A common claim is that the various government quantity and price controls and taxes and subsidies "distort" the efficient use of resources that would be produced by perfect markets. However, markets are never perfect, and the "distortion" may be intended because the operation of actual markets is distrusted and the economic patterns they would create are rejected. Other distortions are not intended but occur as by-products of government actions; their effects may overshadow the primary purpose of those actions. For example, interest-rate ceilings on bank loans, which are intended to encourage new investment, often set the price of funds much below the rental values of the real capital the funds can purchase. Low interest rates may be quite unnecessary in the developing countries, where the incentives to invest are so substantial that low interest rates are not required to encourage new investment. Interest-rate ceilings lower the cost of capital for that portion of the business community with access to the funds at the controlled rate. This, in turn, may lead to the choice of more capital-intensive technologies than would be warranted to maximize output, the rate of growth, or the absorption of labor or to satisfy any of the other criteria of appropriateness.

Similarly, government labor regulations (such as minimum-wage laws, requirements on employers for social contributions, or severance pay regulations) may raise the effective price of labor above its real scarcity value in terms of the goals of the economy. That, in turn, will encourage the substitution of other productive resources for labor and will reduce employment incentives. This is another example of the potential, if not actual, conflict between alternative criteria for the appropriateness of technological choice. The objective of the various minimum wage, welfare contribution, and employment regulations is to improve the distribution of income. The immediate side effect, however, is to create incentives to reduce employment. The final result, as a consequence of reduced employment, may be increased inequality in the distribution of income.

A number of other types of government policies designed to guide and control the development process have similar, though more indirect, effects. Government tariff and quota restrictions on imports and export subsidies, for example, may affect the technological choices. As noted in Chapter 4, it has been alleged that foreign-trade policies are prime determinants of the choices of technology and, in particular, of the absorption of labor in manufacturing. Overvaluation of the exchange rate (for example, by discouraging exports) may emphasize industries with only limited employment potential.

The emergence of unemployment as an open and pressing problem in the developing countries has led to government commitments to deal with it directly, and these commitments, in turn, have become part of development ideology. For example, job guarantees have been required to prevent workers from losing their jobs when they must leave work for sickness or to do military service. In some cases, subsidies have been provided for the employment of labor. These policies encourage the inefficient use of labor and the flow of workers from sectors not covered by the regulations to the covered sectors.[6] In practice, that stimulates migration of labor from rural to urban areas. Thus, one result of such policies is the creation of more open unemployment.

International Private Enterprise and Market Influence on Choice of Appropriate Technologies

The significance of international business as a source of knowledge that affects the choice of appropriate technologies has already been noted. At this point, the issue is the character of the technological decisions that are made directly or indirectly under the influence of international business or multinational corporations.

The primary objective of the multinational corporations, like other private enterprises, must be the long-run maximization of their profits. However, that is not necessarily inconsistent with all the goals embodied in the various criteria for the choice of appropriate technology. Neither is it certain that there is consistency with any of the criteria. The extent to which profit maximization by the multinationals is congruent with national-development goals will depend, in large part, on the character of the regulations and incentives to which they are subject. Yet there are some goals and technological decision criteria that in a range of likely circumstances will be either irrelevant or antithetical to the private-profit objectives of the multinational, as well as to national, firms. For example, no private enterprise, national or multinational, is interested in maximizing employment for its own sake or reducing inequality in the distribution of income for the sake of social welfare, even though their operations may contribute to employment and, possibly, to income equalization under certain circumstances. If objectives of this type are imposed to such an extent that a multinational's relative advantage of operating in a developing country disappears, then the multinationals will simply not invest, and whatever potential contributions they might make will not be realized.

Although private national and multinational corporations both may have the same objective of maximizing their profits, pursuit of that goal by each may well lead to different technological decisions, even in the production of the same commodity and in response to the same set of input prices. Such differences may even be optimal, not only for the individual firm but for the economy as a whole. On the other hand, other differences between the manner in which national and multinational enterprises operate may not be either rational or in the interests of the developing countries.

One difference between national and multinational firms that may indeed be rational and optimal lies in the way they choose between existing and new technologies. National firms may try to develop technologies especially suited to local conditions. But multinationals may prefer to transfer to the developing countries the technologies that have been worked out for the relatively industrialized countries. Such transfers may be undertaken without considering the possibility of developing alternative production methods. However, the decisions may also reflect, implicitly if not explicitly, the advantages of working with well-known methods. To develop techniques more suitable to the labor and other resources of the developing countries, research expenditures and, perhaps, pilot plants and production experience are necessary. By comparison, familiarity with conventional methods makes their implementation less costly and faster, and the outcome more certain. For the multinational corporation, the choice involves comparing the costs of existing methods with all the costs—research and development, investment and operating—of finding and implementing new technologies. In particular cases, the cost comparison can easily favor implementing the known technologies. That decision may also be optimal for the developing country if any plant that gets a new technology has to bear all the costs of research and experimentation.

Their intimate knowledge of production methods may give the multinational corporations an advantage in the use of "second-hand" equipment from their own or other establishments in advanced countries. While not necessarily "appropriate," such equipment may embody relatively labor-intensive technologies especially suited for some of the developing countries. However, the quality of second-hand equipment varies greatly. Thus, since multinational corporations can be expected to be familiar with the used machinery's previous production history, they can more effectively use this means of adapting technology to the conditions of the less-developed countries.[7] Yet this advantage is lost if, as is sometimes the case, the importation of second-hand equipment is restricted.

In general, the technological options for national private or public enterprises of the developing country are different from those for multinationals. For the former, the conventional and well-proved technologies may not be readily accessible or would become available only at a cost, in terms of royalty and license fees, which reflects some monopoly power on the part of the multinational. Moreover, national private and public enterprises in the developing countries are likely to have a longer time horizon in the country than the multinational corporation, so any operating-cost advantages of a new method can be expected to pay off over a longer period. The importance of the latter consideration is reduced by the relatively high discount rate appropriate to investment in the developing countries, but it is not eliminated.

Another source of difference in the choice of technologies by multinational corporations is the prices they pay for their productive inputs as compared to the prices national enterprises pay. Multinational firms borrowing in international finarcial markets may procure credit at lower interest rates than national firms without access to those markets. It has also been argued that multinational corporations investing in a developing country will often raise a significant portion of their funds in the country itself and at a lower interest rate than is available to national enterprise.[8] It may well be rational for national financial institutions to lend to multinational corporations at lower rates than are available to national enterprise. Since the multinational corporation is typically larger than most national firms and may represent a diversified group of enterprises, it is less likely to default than the smaller and less-diversified national enterprises. Moreover, loans by national financial institutions to multinational enterprise can represent a desirable diversification to their portfolios so, again, they may be relatively willing to supply funds. For these reasons, the real cost of financing the multinationals from national sources is likely to be less than the cost of financing national private enterprise, unless national enterprises are subsidized by their government or permitted to borrow at low regulated rates. In such cases, the only advantage left to the multinational is the lower risk of lending to it. But that may well provide enough justification for the national financial institutions to ration credit to satisfy the multinational's loan requests first.

On the other hand, multinationals may well face higher labor costs than national firms. First of all, labor legislation that effectively raises labor costs often does not apply to the many relatively small firms typical of developing countries. Secondly, labor legislation may be differentially enforced among the firms to which it applies in principle. Since the multinationals are relatively conspicuous because of their size and because they are foreign, enforcement for them is likely to be stringent. For analogous reasons, labor unions are likely to be relatively effective in organizing and gaining wage increases from multinational corporations.[9]

If the price of capital is relatively low and the wages of labor are relatively high for multinationals operating in developing countries, they will be induced to adopt relatively capital-intensive methods when choices are possible. Their choice is not the result of any intention to do a disservice to the developing countries. It simply emerges as they try to maximize their profits by finding least-cost methods. For these reasons, and because the technologies used in the industrialized countries are more easily available to multinationals, it is not surprising to find that these are the technologies the multinationals often transfer to the developing countries. Even so, as usual, the picture is mixed; there is evidence that, in some cases, the multinationals have been more assiduous in attempting to adapt their technologies to local conditions than have national firms. A comparative study of can-making techniques in Africa and Thailand showed that multinationals can be both more and less sensitive to local conditions than indigenous firms.[10]

However, there are also irrational and misguided reasons for the transfer to the developing countries by the multinationals of technologies that are inappropriate by any standards. The unwarranted "engineering" preference for modern techniques may especially characterize the multinationals whose knowledge of such techniques is more intimate. The multinationals may also use the royalty payments on their imported technologies as a means of repatriating a larger proportion of their profits than would otherwise be permitted by law. They may insist on using particular technologies to help ensure a market for their intermediate products or their spare parts; or they may justify the choice of their conventional technology as a means of maintaining quality standards, but again those quality standards may be misguided when applied to the developing countries.[11]

Multinational corporations have been criticized as proponents of the dependent roles that the industrialized countries have forced or induced the developing countries to accept. This dependency, it is claimed, deprives the developing countries of autonomy in determining their national policies, since employment, production, and resource-exploitation decisions are made in the context of maximizing the profit of institutions outside the developing country. Thus, it has been argued that the technological decisions made by multinational corporations are also part of the overall climate of dependency.[12]

The "dependency" arguments contain threads of truth that may not be generalizable, arguments for causation that have not been conclusively demonstrated and, above all, a strong assertion of national autonomy in virtually all aspects of economic life as a value that cannot be questioned. That the dependency arguments have some validity is demonstrated, for example, by the multinational corporations' admissions that they have used their relative financial power to gain advantages not necessarily available to national firms. That is certainly an invasion of national autonomy. When this use of financial power takes the form of illegal bribes, national policy is violated in the same sense that illegal acts are violations.

It is pointless, however, to criticize the multinationals because they do not try to advance the economic and political goals of the developing countries. That is not their function, just as it is not the function of national private enterprise. While multinationals may not choose the technologies most suitable for the developing countries, their choices are made to advance their own self-interest rather than in conspiracy against the interests of the developing countries. In advancing their interests, they can be expected to take advantage of whatever monopoly power and economic and political influence they command.

Careful regulation can prevent the multinationals from using their influence in directions counter to national goals. But that, in turn, requires a government capable of overseeing and influencing national economic processes. Creating such capabilities is one of the central problems of political development and is by no means an automatic and easy achievement of national governments.

Influence of Bilateral and Multilateral Official Institutions on Choice of Appropriate Technologies

Through their official relations the governments of the industrialized countries and the international economic assistance institutions provide technological information to developing countries in the course of sponsoring particular projects. In addition, they also exert a broad influence over the economic policies of recipient nations, including their policies with respect to technological decisions. This influence is contained in part in the terms imposed on the developing countries as conditions of loans or grants from governments and international institutions. That the influence is not always exerted exactly in the directions the developing countries themselves would prefer is demonstrated by their complaints about the "strings" and "leverage" accompanying official international loans and grants. While the character of this leverage has been the subject of considerable controversy, it is now virtually officially acknowledged in general, though little, if anything, is publicly known in detail.[13]

To ameliorate the impact of development assistance on the donors' balance of payments, most of such assistance provided through bilateral arrangements is "tied" to purchases within the lending government's economy. As a result, the range of technological alternatives from which choices may be made is restricted, and that can lead to inappropriate choices. When loans and grants are tied to expenditures in the lending/granting nation, the assistance may have to be used in a relatively high-cost market. This was the effect of tying United States assistance before the 1971 depreciation of the dollar. In effect, such tying to a relatively expensive market reduces the value of the economic assistance that is provided.[14] The developing countries may decide to choose relatively capital-intensive methods because those are the only ones available. However, though the methods appear capital intensive, they may not be, in reality, since no assistance would be provided if the particular techniques were not used. The real cost to the developing country is not the "face value" of the equipment but something substantially less. The casual observer, seeing only the automated textile factory resulting from assistance, could well come to the wrong conclusion.

The precise character of the influence that official sources have on the technological decisions of the developing countries will vary with individual government assistance agencies and multinational lending agencies, and the manner in which the influences manifest themselves will also differ among the developing countries. The assistance programs of the industrialized countries have a variety of rationalizations, ranging from humanitarian to the narrowest self-interest. Critics of dependency have charged that the industrialized countries have consciously used assistance programs to maintain the developing nations in a technologically subservient position by refusing to supply advanced technology, by supplying only obsolete technology, or by not supplying the technology and loans needed to develop sectors that would increase the self-sufficiency of the developing nations.[15] Without more information than is available on loan and grant negotiations and influences on technical choices, these arguments cannot be brushed aside. On the other hand, a broad range of industries and technologies has been supported by assistance programs, including, for example, the most modern high-speed computer and petrochemical plants as well as technical assistance in simple agricultural and handicraft production. This, in itself, suggests that there is at least no overall and systematic pattern of technological discrimination.

All of the international agencies are subject to constraints that, to some extent, prescribe their lending terms. For example, the International Bank for Reconstruction and Development (World Bank) borrows in international financial markets a substantial part of the funds it lends. It is understandable, therefore, that the criteria the World Bank applies to determine the use of such funds are similar to those used by private enterprise. These, as noted above, can be narrower than the criteria that policy makers in developing countries might use. A substantial part of the World Bank's funds are subscribed by its member governments, and governments that are the major sources of funds and shapers of World Bank policy in the use of these funds may also have different criteria for development from those of the developing nations.

It is also clear that the assistance agencies of the industrialized countries and the international lending institutions often have a different conception of the most promising growth policies than do the developing nations themselves.

This is clearly true of the World Bank, as example and epitome of international lending agencies. Its conceptions change over time, however, under the impact of the arguments of the developing countries and new perceptions of the development process. Only recently, for example, has the World Bank recognized employment, income distribution, and rural development as worthy development goals different from growth in output and balance of payments correction. Whatever the sources, the conceptions of development processes held by such agencies as the World Bank provide the intellectual basis for the leverage that, according to authoritative sources, is exercised in giving their assistance.[16]

The International Monetary Fund and other international lending agencies have received less critical attention than the World Bank. Yet, there is evidence that each also has its characteristic views about the national development policies that are worthy of its assistance.[17]

Once it is accepted that these national and international economic assistance agencies do use their power to influence the policies of developing countries, the question that then arises is what determines the policy choices of these agencies and, in particular, their criteria for appropriate technologies. Presumably, the agencies would answer that their preferences reflect the study and judgment of their staffs, distilling whatever wisdom is available in the various relevant professions. Unfortunately, such an answer is not conclusive, as there are strong differences of opinion concerning the validity of the agencies' positions. Not only may each developing country claim that its circumstances are unique; its interpretation of development theory and experience may also lead to different conclusions. Moreover, they can well claim that the bilateral and multilateral assistance agencies themselves have self-contradictory rules and goals. For example, it is common now for projects being considered by national or multilateral assistance agencies to be judged on the basis of a well-defined system of cost–benefit analysis whose criterion may be social cost minimization, consumption maximization, or growth. That might point to a particular source of supply of machinery and technological information. Yet, as noted previously, the tying of assistance by national agencies may force recipients to use less desirable technologies than would otherwise be chosen. For example, although investment decisions in a thermoelectric station depend on a cost–benefit analysis, the choice of equipment may be confined to relatively high-cost suppliers in countries supplying the foreign exchange. In addition, the developing countries can be expected to have different development goals than those set for them by assistance agencies. These differences manifest themselves in the risks they are willing to take with respect to inflation or balance of payments problems or other development difficulties.

REFERENCES AND NOTES

1. For a classic argument see P. Rosenstein-Rodan (1943).
2. The theoretical analysis of the implications of competitive systems for the personal distribution of income has, as yet, produced results that, at best, are asymptotic with little indication as to rate of progress toward the asymptote. See J. Stiglitz (1969), K. Meesook (1974), and A. Dasgupta (1975).
3. H. Hughes (1974) and G. Ranis (1974).
4. L. Wells, p. 83, in P. Timmer et al. (1975).
5. E. Edwards (1974), p. 17.
6. See, for example, J. Harris and M. Todaro (1970).
7. The comparative study of can-making techniques used in Kenya, Tanzania, and Thailand of C. Cooper, R. Kaplinsky, R. Bell, and W. Satyarakevit, in A. Bhalla (1975), provides interesting insights into the variety of types of behavior of multinational firms. In some cases it is quite sensitive to local factor prices, in other cases it is not.
8. For contrasting views see, for example, W. Manser (1974) and S. Plasschaert (1974).
9. For a strikingly implicit argument see N. Weinberg, in A. Said and L. Simmons (1975).
10. C. Cooper, R. Kaplinsky, R. Bell and W. Satyarakevit, in A. Bhalla (1975).
11. For a strong view see H. Mowlana, in A. Said and L. Simmons (1975).
12. O. Sunkel (1973).
13. T. Hayter (1971) contains descriptions of the leverage applied by the International Bank for Reconstruction and Development in its lending operations and allegations of attempts to suppress their publication. This leverage was later mentioned rather briefly in the "official" history of the World Bank by E. Mason and R. Asher (1973).
14. See J. Bhagwati (1970).
15. See T. Hayter (1966).
16. E. Mason and R. Asher (1973) provide a detailed description of the operations of the World Bank to 1973 without, however, an analysis of the rationale.

6 Special Features of Technological Decisions in Agriculture

Introduction

After years of receiving little analytical and administrative attention and relatively small investment allocations, the central role of the agricultural sector in development has become more widely appreciated. The relative size of the agricultural sector in the developing countries itself warrants particular attention to the sector. In most of the developing countries, the share of agricultural output in total output ranges between 30 and 50 percent and in some cases is even larger. Since the earnings of labor are generally lower in agriculture than in industry and government, the proportion of the work force in agriculture is higher than the share of output. This means that relatively small changes in the demand for labor in agriculture result in major changes in the availability of labor to the urban manufacturing and service sectors. Suppose, for example, that the total labor force of a developing nation were set at 100, with 60 in agriculture, 20 in industry, and 20 in services. If there were a 2-percent annual increase in labor productivity in agriculture with no opportunities for increased employment in that sector, then 1.2 laborers would be released per year from agriculture. These would have to be absorbed by the other sectors if unemployment were not to increase. That would mean an annual increase in employment in both industry and services of 3 percent. If only manufacturing is expanding, employment there would have to grow by 6 percent. But if labor productivity in the nonagricultural and other sectors were also growing, as would be expected in the course of development, then output in these other sectors would have to grow still more rapidly to absorb all the labor displaced from agriculture.

In addition to its size, the other unique features of agriculture in most of the developing countries justify an analysis specially tailored to it. Because of its unique characteristics, agriculture cannot simply be "left to itself" if development is not to be frustrated by agricultural scarcities, if urban unemployment problems are not to overwhelm the economy, and if all sectors are to participate equitably in the benefits of economic growth. This chapter will identify and describe briefly the unique features of agriculture and relate them to the special problems of finding and implementing appropriate technologies in the sector.

Diversity in Agriculture

The great diversity in the technical conditions of production in agriculture has several sources. The products of agriculture are themselves quite heterogeneous and have diverse input patterns. The requirements for the production of the row crops of grains or legumes are different from the requirements for tree crops of fruits, nuts, berries, or leaves and from the requirements needed to raise animals. Some of the inputs needed to grow staple grains, for example, may be completely irrelevant for raising livestock, and the land and water requirements can be quite different. The products also differ in the degree to which they can be modified and adjusted to diverse geographic conditions. Land and climate conditions, of course, have great effects on productivity of labor and other resources employed in agriculture. While labor and capital can substitute for unfavorable natural conditions to some degree, the extent and ease of substitution depends on natural conditions.

The limits on our knowledge make it impossible to say that the range of physically efficient input proportions is greater in agriculture than in other sectors. Such comparison is difficult not only because knowledge of the technology in other sectors is limited, but also because even in agriculture it is seldom possible to know in detail the range of input proportions suitable for each locale. Nonetheless, the prevailing impression is that there is considerable variety in potentially economical input choices for important types of agricultural outputs.[1] Figure 1, which compares the per-acre use of fertilizer and per-worker use of tractor horsepower among countries, illustrates for these two factors the wide range of intensity of use of resources in agriculture. Even within the same locale, the range of technological choices may be different for different products.

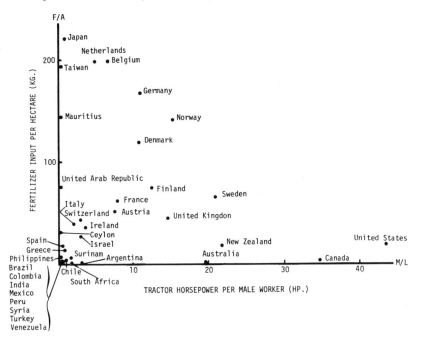

FIGURE 1 International comparison of fertilizer per hectare and tractor horsepower per male worker. From Hayami and Ruttan (1971), p. 72. Data on fertilizer use are 1957–62 averages, and tractor data are for years closest to 1960.

Because the range of efficient resource inputs in agriculture is wide, scheduling tasks in diversified farming can be quite complex. Yet peasant farming may have achieved a high degree of management efficiency. The requirements of each type of activity are well-known, at least as that activity has been practiced, and the trial-and-error methods of farmers might be an effective way of arriving at optimal scheduling.

Another source of the diversity in the developing countries' agricultural technologies is the wide geographic origin of many strains of plants and animals. A crop grown by an apparently traditional and otherwise isolated peasantry may be using seed generated in an experiment station on the other side of the world. Analogously, the traditional farmer may be using irrigation water flowing in ditches from a distant dam, which reflects very different technologies from that being used in peasant farming.

Associated with the diversity of production conditions is a remarkable diversity in the markets in which agricultural products are sold.

Some products are highly local, because they cater to local tastes or because they are perishable. Often perishability will depend on the available transportation and marketing arrangements. Other agricultural products are grown specifically or mainly for international trade and move long distances. They may be marketed in economies much more or, perhaps, less advanced than those in which they originate. And other commodities can move in or out of local or national and international markets depending on demand and supply conditions.[3]

These differences in market conditions mean that the agricultural technologies used must face, with different degrees of stringency, the tests of output and cost. If production is undertaken only for local markets, it does not have to compete directly with internationally marketed products. Therefore, the technologies used for locally marketed products may more readily be chosen for characteristics other than cost. However, for products subject to international competition, either the cost minimization criterion of appropriateness must be adopted or the government must stand ready to use its tax and subsidy powers to preserve technologies adopted on other criteria.

Agricultural organization in the developing countries also varies greatly. There are latifundia and minifundia, permanent and temporary tenancies, sharecropping and freeholding, and each of these forms and others have special features in each country, even in regions within each country. This diversity reflects to a considerable degree the influence of local traditions and social patterns in agriculture. It is in this sector that the influences of each country's unique patterns of culture and social structure appear to be preserved longest and have the greatest weight.

While there may be little controversy over these general references to the diverse organizational structures found in the agricultural sector, there is a substantial controversy over the implications these structures have for the choice of technologies and the use of resources. As noted earlier, one school of thought has argued that traditional agriculture uses resources efficiently, both physically and economically. The relative poverty associated with much of traditional agriculture is then explained as the result of the limited productive capabilities of its technology and/or the high cost of capital to the agricultural sector.

On the other hand, another view holds that various tenancy and sharecropping arrangements and the landlord monopoly power important in the agricultural sectors of developing countries diminish efficiency in the use of resources and lead to the choice of inappropriate technologies.[4] And, as pointed out earlier, the treating of family labor as a fixed resource in family farming can itself lead to inefficiencies in the use of all resources.

Unfortunately, a definitive resolution of these differing views is not possible at present. However, implicit in many policy recommendations made for developing countries' agricultural sectors is that a policy of *laissez faire* is not appropriate. Yet it would be if the sector fit the model of perfect competition and operated efficiently. That agriculture is not believed to fit this model is evident, for example, in the arguments that claim large landholders have access to credit, fertilizer, and even government technical assistance on specially favorable terms.[5]

Sources of Knowledge and Technological Change in Agriculture

There is a long history of local experimentation leading to technological change in agriculture. Many stories of landowners with a scientific bent trying to improve local agricultural methods and to propagate those improvements testify to this pattern. And the continuing eagerness of some farmers to adopt and demonstrate new innovations indicates its persistence. In the typical story, the experimenting and innovating farmer is a relatively large landowner who can better risk failure than the small farmer. However, the dramatic innovations in agriculture have more and more been the kind that could not be generated locally and on the farm. For the most part, improvements in seed quality, tractorization, irrigation, fertilizer, and pesticide use have not originated locally, though there may continue to be some innovations like these at the local level.[6]

A factor discouraging individual technological experimentation in agriculture is the impossibility in many cases for the single entrepreneur to capture the full benefits of his innovation. It is no accident that hybrid seed production is a field in which private entrepreneurship has been relatively successful, because that seed is not generated directly by the crops it produces. That permits private seed producers to capture at least part of the benefits of the higher productivity of hybrid seeds.

There are other reasons that research on technological change in agriculture is increasingly centralized, though, as will be noted, not without the influence of local conditions. For one, the "technology" necessary to produce modern innovations is relatively large in scale, has long time horizons, and is science intensive, particularly for biological innovations. The development of many of the new strains of staple grain crops and of sugarcane illustrates the extended periods of time, the large scale, and the dispersed efforts needed for these innovations.[7] Seeds and plants were assembled from many different sources, and new strains evolved over a number of years under the direction of highly specialized plant scientists to achieve the varieties that have currently demonstrated their superiority. This has been followed by a relatively long period of local adaptation of new technologies, which again often requires a relatively large-scale effort and long time horizon. By comparison, the typical scale of agriculture in developing countries is small, and even the largest landlords could seldom support the experiment stations and production facilities necessary for most modern innovations. Thus, individual farmers simply cannot undertake the research and development necessary to develop the full range of new technologies in agriculture.

Technological change in the agriculture of the developing countries was originally a matter of importing new seeds, animals, and equipment from industrialized countries. While this continues to some extent, the methods of plant science, animal husbandry, and machine design are now also transmitted and used to develop agricultural materials suited for local conditions in the developing countries.[8]

There have been some remarkable successes in agricultural research, in particular in the development of high-yielding wheat and rice strains at the International Rice Research Institute (IRRI) and the International Center for the Improvement of Maize and Wheat (CIMMYT). Table 2 summarizes the income and costs of research in these two organizations. There can be little doubt that the achievements of CIMMYT and IRRI have helped stimulate the formation of other international agricultural research organizations. Table 3 lists 12 such organizations; 2 of these are still in the planning stages.

While the successes of IRRI and CIMMYT are undeniable, it is not yet certain that the other international research institutes can reproduce such successes, nor is there complete agreement on the conditions that permitted these initial successes. However, there does appear to be considerable agreement that future success in the local adaptation of new agricultural technologies will require national research institutes of high quality. Indeed, it has been argued that effective local adaptation will require national institutes whose quality is like that of the international research institutes and which, therefore, could also make original contributions to scientific and technical knowledge.[9] The training of personnel from the developing countries has succeeded in creating capacity in some of them for the generation of completely indigenous agricultural techniques and material.

The technological innovations in agriculture cannot be assumed to have uniform effects in widening or narrowing the range of efficient alternatives. In some cases, they may generate a new set or range of input proportions that simply dominate all the preexisting methods by using all types of resources more effectively. The new technologies

TABLE 2   Income Stream and Cost Calculation for International
Center Research[a]

Item	First-Generation Varieties, 1965–6 to 1969–70		Second-Generation Varieties, 1970–1 to 1973–73	
	Wheat	Rice	Wheat	Rice
Annual increment to income stream (1973, million $)	135	270	56	360
Associated cost on an annual basis[b] (1973, million $)	0.6	1.0	1.2	2.8
Income stream per $1,000 investment	$225,000	$270,000	$46,666	$128,500

[a] REFERENCE: Thomas M. Arndt and Vernon W. Ruttan, *Conference on Resource Allocation and Productivity in National and International Agricultural Research*, A Seminar Report (New York: The Agricultural Development Council Inc., September 1975).

[b] Computed from Dalrymple (1975), Table 1. The second-generation costs are based on annual budgets for 1966–68. First-generation costs are all prior costs at IRRI (capital expenditures are amortized) and a capital adjustment is made for CIMMYT costs to make them roughly comparable with IRRI.

SOURCE: Robert E. Evenson, "Comparative Evidence on Returns to Investment in National and International Agricultural Research," ADC/RTN Conference on Resource Allocation and Productivity in National and International Agricultural Research, Airlie House, Virginia, January 26–29, 1975.

may, however, greatly change the productivity of one or another
resource. The high-yielding varieties of wheat and rice in some circumstances are "land augmenting" because their shorter growing season
makes it possible to grow another crop on the same land. Tractorization can, in some circumstances, have the same effect if it breaks a
bottleneck at some critical phase of a growing season. But these
innovations also increase the productivity of labor as well. The net
"bias" in the innovations is not readily determined.[10]

It has been claimed that an "induced innovation" process in agriculture has worked in advanced countries and can work in the developing
countries as well.[11] According to this argument, the needs of private
farmers stimulate the technological changes even though such changes
are carried out in large-scale public research institutions far from the
farmers themselves. The argument postulates that the personnel of the
research institutions are sensitive to farm profitability as determined by
product prices and input costs and that organized farmers, the incentives of professional recognition, and the assistance of efficient private
agricultural supply firms stimulate the research personnel to respond to
the farmers' needs.

That research institutes respond in this way can hardly be doubted

TABLE 3 Present Structure of the International Agricultural Research Network

Center	Location	Research	Coverage	Date of Initiation	Proposed Budget for 1975 ($000)
IRRI (International Rice Research Institute)	Los Banos, Philippines	Rice under irrigation; multiple cropping systems; upland rice	Worldwide, special emphasis in Asia	1959	8,520
CIMMYT (International Center for the Improvement of Maize and Wheat)	El Batan, Mexico	Wheat (also triticale, barley); maize	Worldwide	1964	6,834
CIAT (International Center for Tropical Agriculture)	Palmira, Colombia	Beef; cassava; field beans; farming systems; swine (minor); maize and rice (regional relay stations to CIMMYT and IRRI)	Worldwide in lowland tropics, special emphasis in Latin America	1968	5,828
IITA (International Institute of Tropical Agriculture)	Ibadan, Nigeria	Farming systems; cereals (rice and maize as regional relay stations for IRRI and CIMMYT); grain legume (cowpeas, soybeans, lima beans, pigeon peas); root and tuber crops (cassava, sweet potatoes, yams)	Worldwide in lowland tropics, special emphasis in Africa	1965	7,746
CIP (International Potato Center)	Lima, Peru	Potatoes (for both tropics and temperate regions)	Worldwide including linkages with developed countries	1972	2,403

ICRISAT (International Crops Research Institute for the Semi-Arid Tropics)	Hyderabad, India	Sorghum; pearl millet; pigeon peas; chick-peas; farming systems; groundnuts	Worldwide, special emphasis on dry semiarid tropics, nonirrigated farming. Special relay stations in Africa under negotiation	1972	10,250
ILRAD (International Laboratory for Research on Animal Diseases)	Nairobi, Kenya	Trypanosomiasis; theileriasis (mainly east coast fever)	Africa	1974	2,170
ILCA (International Livestock Center for Africa)	Addis Ababa, Ethiopia	Livestock production systems	Major ecological regions in tropical zones of Africa	1974	1,885
IBPGR (International Board for Plant Genetic Resources)	FAO, Rome, Italy	Conservation of plant genetic material with special reference to cereals	Worldwide	1973	555
WARDA (West African Rice Development Association)	Monrovia, Liberia	Regional cooperative effort in adaptive rice research among 13 nations with IITA and IRRI support	West Africa	1971	575
ICARDA (International Center for Agricultural Research in Dry Areas)	Lebanon	Probably a center or centers for crop and mixed farming systems research, with a focus on sheep, barley, wheat, and lentils	Worldwide, emphasis on the semiarid winter rainfall zone		
INTSOY (Soybean Improvement)					1,900 (1977)

SOURCE: Nicholas Wade, "International Agricultural Research," *Science*, vol. 188, pp. 585–589, Table 1, 9 May 1975, and Robert E. Evenson and Yoav Kislev, *Agricultural Research and Productivity* (New Haven: Yale University Press, 1975), p. 28.

from the descriptions of the processes by which some of the important agricultural innovations have occurred. However, the analysis thus far provided does not appear to define rigorously the processes of adjustment that might make the process operate efficaciously.

The attempts to measure the productivity of research in agriculture have generated some impressive rates of return. The results of some studies are summarized in Tables 4 and 5. Yet the estimates themselves are the subject of considerable controversy. There are claims that they overestimate the revenues attributable to research and underestimate the costs and counterarguments that these numbers really underestimate the returns.[12]

In some cases, concentrating agricultural planning and research activities in institutions separate from farmers has in fact led to inappropriate technologies and the production of inappropriate products. For example, the tractorization of agriculture in the Soviet Union, depending on relatively large machines, is commonly regarded in the West as a mistake.[13] It is easy to treat this experience as an inevitable result of allowing ideology to intrude into what should be

TABLE 4 Summary of Direct Cost-Benefit-Type Studies of Agricultural Research Productivity[a]

Study	Country	Commodity	Time Period	Annual Internal Rate of Return (%)
Griliches (1958)	U.S.A.	Hybrid corn	1940–1955	35–40
Griliches (1958)	U.S.A.	Hybrid sorghum	1940–1957	(20)
Peterson (1966)	U.S.A.	Poultry	1915–1960	21–25
Ardito–Barletta (1970)	Mexico	Wheat	1943–1963	90
Ardito–Barletta (1970)	Mexico	Maize	1943–1963	35
Evenson (1969)	S. Africa	Sugarcane	1945–1962	40
Ayer (1970)	Brazil	Cotton	1924–1967	77+
Hertford, Ardila,	Colombia	Rice	1957–1972	60–82
Roches and	Colombia	Soybeans	1960–1971	79–96
Trujillo (1975)[b]	Colombia	Wheat	1953–1973	11–12
	Colombia	Cotton	1953–1972	none
Peterson and	U.S.A.	Aggregate	1937–1942	50
Fitzharris (1975)[b]			1947–1952	51
			1957–1962	49
			1967–1972	34

[a] REFERENCE: Thomas M. Arndt and Vernon W. Ruttan, *Conference on Resource Allocation and Productivity in National and International Agricultural Research,* A Seminar Report (New York: The Agricultural Development Council, Inc., September 1975).

rational calculation. But ideologies, not only political, always have influence when knowledge is less than complete, as is necessarily the case with agricultural production conditions. Another example of the potential for mistakes in agricultural innovations is the famous "groundnut" scheme of the early post-World War II period. The attempt to establish the large-scale cultivation of peanuts in Tanzania ended in complete failure after an expenditure of roughly $100 million.[14]

It has also been argued that the Green Revolution as a whole embodies many mistakes, if not just of technology then of inadequate appreciation of its wider implications.[15] Agricultural yields undoubtedly increased dramatically when high-yielding varieties of wheat and rice seeds were introduced. However, a number of unforeseen problems arose as the new seeds were dispersed. For one, concern has emerged that the Green Revolution might contribute to labor displacement and concentration of income, but these outcomes do not seem to be as big a problem as some originally feared.[16]

Until recently, the explanations and discussions of the technological

---

[b] From papers presented at ADC/RTN Conference on Resource Allocation and Productivity in National and International Agricultural Research, Airlie House, Virginia, January 26–29, 1975.

SOURCES: The estimates that were presented at the Conference on Resource Allocation and Productivity in National and International Agricultural Research (Airlie House, Virginia, January 1975) are identified. The other estimates have been summarized by James K. Boyce and Robert E. Evenson, *National and International Agricultural Research and Extension Programs* (New York: The Agricultural Development Council, Inc., August 1975).

The sources of the individual estimates are as follows:

Ardito-Barletta, N. *Costs and Social Returns of Agricultural Research in Mexico,* Ph.D. Dissertation, University of Chicago, 1970.

Ayer, H. *The Costs, Returns and Effects of Agricultural Research in a Developing Country: The Case of Cotton Seed Research in Sao Paulo, Brazil,* Ph.D. Dissertation, Purdue University, 1970.

Evenson, R. *International Transmission of Technology in Sugarcane Production,* Mimeo, Yale University, 1969.

Evenson, R. *The Contribution of Agricultural Research and Extension to Agricultural Production,* Ph.D. Dissertation, University of Chicago, 1968.

Griliches, Z. "Research Costs and Social Returns: Hybrid Corn and Related Innovations," *Journal of Political Economy* 66:419–431, 1958.

Griliches, Z. "Research Expenditures, Education and the Aggregate Agricultural Production Function." *American Economic Review,* December 1974.

Hertford, R.; Ardila, J.; Roches, A.; and Trujillo, C. "Productivity of Agricultural Research in Colombia." ADC/RTN Conference on Resource Allocation and Productivity in National and International Agricultural Research, Airlie House, Virginia, January 26–29, 1975.

Hines, J. *The Utilization of Research for Development: Two Case Studies in Rural Modernization and Agriculture in Peru,* Ph.D. Dissertation, Princeton University, 1972.

Latimer, R. *Some Economic Aspects of Agricultural Research and Extension in the U.S.,* Ph.D. Dissertation, Purdue University, 1964.

Kahlon, A.S.; Bal, H.K.; Saxena, P.N.; and Jha, D. "Productivity of Agricultural Research in India," ADC/RTN Conference on Allocation and Productivity in National and International Agricultural Research, Airlie House, Virginia, January 26–29, 1975.

Peterson, W.L. *Returns to Poultry Research in the US,* Ph.D. Dissertation, University of Chicago, 1966.

Peterson, W.L.; and Fitzharris, J.C. "Productivity of Agricultural Research in the United States," ADC/RTN Conference on Resource Allocation in National and International Agricultural Research, Airlie House, Virginia, January 26–29, 1975.

Tang, A. "Research and Education in Japanese Agricultural Development," *Economic Studies Quarterly,* 1963.

TABLE 5 Summary of Selected Sources-of-Growth-Type Studies of Agricultural Research Productivity[a]

Study	Country (Commodity) Time Period	Annual Internal Rate of Return (%)
Griliches (1964)	U.S.A. (Aggregate) 1949–1959	35–40
Latimore (1964)	U.S.A. (Aggregate) 1949–1959	Not Significant
Evenson (1969)	U.S.A. (Aggregate) 1949–1959	47
Tang (1963)	Japan (Aggregate) 1880–1938	35
Ardito-Barletta (1970)	Mexico (Crops) 1943–1963	45–93
Peterson (1966)	U.S.A. (Poultry)	21
Evenson (1969)	South Africa (Sugarcane) 1945–1958	40
Evenson (1969)	Australia (Sugarcane) 1945–1958	50
Evenson (1969)	India (Sugarcane) 1945–1958	60
Jha and Evenson (1973)	India (Aggregate)	40
Kahlon, Bal, Saxena, and Jha (1975)[b]	India (Aggregate)	63

[a] REFERENCE: Thomas M. Arndt and Vernon W. Ruttan, *Resource Allocation and Productivity in National and International Agricultural Research,* A Seminar Report (New York: The Agricultural Development Council, Inc., September 1975).
[b] From papers presented at ADC/RTN Conference on Resource Allocation and Productivity in National and International Agricultural Research, Airlie House, Virginia, January 26–29, 1975.
SOURCES: See Table 4.

issues in agricultural innovation showed little evidence that criteria of appropriateness other than increasing output with available resources had been considered. For the most part, it has been presumed that increases in output will make it possible to achieve other desired goals, and the possible contradictions among goals has seldom been seriously considered. This focus has changed with the introduction of the high-yielding varieties of wheat and rice seeds. These innovations have had such far-reaching effects that, at least after the fact, they have directed a great deal of attention to the issues emphasized in other appropriateness criteria. The international agricultural research institutes are responding to these issues by taking on a wide range of responsibilities for agricultural development, though this pattern of evolution has been questioned.[17]

It is only fair to note, however, that agricultural colleges and research institutions have a long history of concern with all aspects of agricultural life. These concerns have been compartmentalized in specializations on plant science, animal husbandry, marketing, land reform, agricultural finance, and so on. However, the agricultural extension services have often taken a comprehensive approach to adaptations to new technologies if not in the original decisions.

Policies for Implementation of Technological Decisions in Agriculture

In most of the developing countries, the agricultural sector is primarily or exclusively characterized by private enterprise. Collective and cooperative farms are found in some of these countries, but these are still unusual. Private enterprise in agriculture takes many different forms in the developing countries, however. Some is "capitalist" in the sense that all labor and other resources used are hired and purchased, and technology is chosen to maximize the profits of the enterprises. Although family enterprises may hire and buy some labor and resources, for many family farms the family labor is, to a considerable degree, a "fixed resource." Family resources will be used to maximize the family's satisfaction from output, from leisure, and from consumption. Just how that is determined will vary with country and culture. But it cannot be assumed that profits, in the customary sense, will be maximized in farms on which most of the labor is supplied by the family.

It is a disservice to the cultural and social richness of the rural areas of the developing countries to cast them all in the mold of perfect competition. There is enormous variety, and that variety has important consequences for the choice of appropriate technologies in the agricultural sector. If social and cultural variety itself is considered worth preserving, then technologies should be chosen to avoid giving direct or indirect advantages to new types of organization that will destroy the traditional types. Or, if such technologies are implemented, then traditional forms must be protected by organization, regulation, tax preferences, or subsidies. For example, new technologies commonly have economies of large-scale production that make it impossible for the smallest farmers to compete effectively and survive if they buy and sell in product and resource markets. If these small farmers represent a way of life that is to be preserved, then either the new technologies must be excluded or the small farmers must be protected from their effects or organized to benefit from them.[18]

Like most private enterprise, private farmers have no reason to pursue the overall economic goals of the economy. They are simply not concerned with aggregate economic growth, regional development, balance of payments correction, and political development. So the criteria that private farmers will use to make technological decisions will be individualist. Family, as compared to capitalist, enterprise may be concerned with providing employment for its members and may adjust its technological decisions and use of resources to do this. Interpretations of the quality of life can again be expected to be dominated by the local culture and social structure, though agricultural extension services actively attempt to exercise influence.

However, national policies have been adopted in many countries to force or induce individual farmers to make decisions that are consistent with overall economic goals. Price supports may be used to encourage the production of a particular crop or the development of the agriculture in a particular region. Price ceilings may be imposed to control the cost of food in urban areas; thus, in a sense these are redistributional. Tax policies may also be designed for their redistribution effects. Export subsidies may be used to encourage production that will help with the balance of payments. While direct or indirect controls on prices and output levels have been used to achieve overall goals, the encouragement or discouragement of particular technologies has also had unintended consequences in the use of resources. For example, price controls or export subsidies that favor planting particular crops over others will result in the use of the combinations of land, labor, and other inputs suited particularly to the preferred output.

Direct regulations may also have unintended effects on the cropping decisions and, therefore, on the technologies used. Land-reform programs often make it difficult to achieve consistency among criteria of appropriateness, including the distributional goals of land reform as well as efficiency and cost minimization. Such programs, typically, are intended to achieve distributional objectives. If they are to succeed, the technologies adopted must be consistent with such objectives. Land reforms that limit the size of landholdings will influence the choice of crops and, thus, the technologies used. When the adopted technologies are appropriate for achieving output or growth objectives but also contain incentives not consistent with the land reform objectives, regulations and fiscal policies may be needed to preserve the goals of land reform. Analogous problems arise in the control of fertilizer distribution and water allocations.

In an increasing number of less-developed countries, the diffusion of technological knowledge has been institutionalized in agricultural extension services. This has often been done at the urging of the industrialized countries that provide economic assistance and in which extension has been regarded as a major instrument of technological change in agriculture. The United States has taken a leading role in this type of institutional development. Again, however, there are substantial questions as to whether extension services have ever been successful in diffusing technology in the absence of clear economic incentives and whether these services are necessary when such economic incentives exist.[19]

Two factors make the social and political organization of agriculture particularly important in the adoption of new technologies in some of the developing countries. First, the structure of agricultural society in many developing areas is not strongly individualistic but primarily communal, with the community being defined in terms of caste, religion, village, personal allegiances, or otherwise. Decision making in these circumstances is, to a considerable extent, also done on a communal basis. Secondly, to be efficient, the scale of investment in, and operation of, the new technology, particularly for mechanical equipment, is often larger than the small-scale farms found in many developing countries. As a result, the speed with which new methods are adopted and the way the benefits are distributed depends on how well the small farmers can organize themselves to take advantage of new technologies and capture potential economies of scale. The Farmers Associations of Taiwan, which reportedly include 95 percent of all farmers, provide an impressive example of organizational success that has strongly influenced the spread of efficient technologies. An example of contrasting effectiveness is provided in a study of the adoption of tube wells in Bangladesh and Pakistan.[20] The study showed that in Bangladesh, where there are many small-scale farms, the use of tube wells facilitated the formation of village cooperatives. In Pakistan, however, where there is also a large proportion of small farms, cooperatives were not formed, in large part because of the high degree of factionalism within the villages. In addition, the large landlords used their political power to obtain the subsidized credit for tube wells offered by the government, leaving the small farmer again at a disadvantage. Thus, the differing local social and political patterns determined the way technology was used and how the benefits were distributed.

As these examples suggest, the criteria of output maximization or cost minimization has been dominant in the search for new agricultural technologies. However, their actual adoption will be strongly affected by the other criteria of technological appropriateness and the manner in which the criteria are implemented by government policy.

REFERENCES AND NOTES

1. Research on U.S. agriculture suggests that the elasticity of substitution is greater than one (Z. Griliches, 1964, p. 970) and studies for agriculture in less-developed countries also indicate an elasticity of substitution greater than or equal to one (Y. Hayami and V. Ruttan, 1971, pp. 105–106).
2. For a strong view see T. Schultz (1964), Chap. 3.
3. This diversity is exemplified in many books and articles and is particularly stressed in M. Millikan and D. Hapgood (1967).
4. See, for example, P. Bardhan and T. Srinavasan (1971).
5. "In Pakistan subsidized credit for the purchase of tube wells has gone almost entirely to large farmers." C. Gotsch (1972), p. 333.
6. For a survey and analysis of agricultural research see R. Evenson and Y. Kislev (1975).
7. R. Evenson and Y. Kislev (1975), Chap. 3.
8. For a survey of a recent conference on national and international research in agriculture see T. Arndt and V. Ruttan (1975).
9. T. Arndt and V. Ruttan (1975), p. 9.
10. Y. Hayami and V. Ruttan (1971), p. 291.
11. Y. Hayami and V. Ruttan (1971), p. 191.
12. See T. Arndt and V. Ruttan (1975), p. 5.
13. T. Schultz (1964), p. 123.
14. F. Burke (1965), pp. 34–39.
15. W. Falcon (1970).
16. C. Gotsch (1972), p. 333.
17. T. Arndt and V. Ruttan (1975), p. 11.
18. See C. Gotsch (1972).
19. See, for example, E. Rice (1971).
20. C. Gotsch (1972).

7 Special Sectoral Problems and Opportunities

Introduction

The issues related to the choice of appropriate technologies in small-scale and service establishments deserve special attention because small-scale enterprise occurs in many industries and it often appears to use technologies especially well adapted to local labor and resource supplies. The service industries, particularly construction, are not only important in development but also seem to provide scope for substitution among resource inputs.

No overall survey of the relative importance of small-scale enterprise appears to be available. The definition of such enterprise itself varies among countries. Thus, a rigorous general assessment of the role of small enterprise in developing countries is not possible at this time. Yet, bits of evidence suggest its significance as a form of economic organization for labor use. For example, a study of small-scale firms in Colombia indicated that 12.2 percent of employees in all industries worked for firms hiring less than 10 workers. This would represent an even larger proportion of the total industrial labor force, since the number of unpaid family workers in small firms is relatively high. An earlier study for India reported that about three-fourths of the 16.2 million persons employed in manufacturing were in rural household enterprise, urban household enterprise, and small-power-using factories hiring less than 10 persons or non-power-using factories hiring up to 20 persons.

99

Table 6 presents some summary data with some sectoral detail for the Philippines, Taiwan, South Korea, Thailand, and Singapore for the 1960s showing the proportion of employment in firms engaging less than 10 persons. From this data emerges an even stronger impression of the significance of small firms.

There is also evidence that small firms use labor more intensively relative to capital and output than do large firms. This suggests that expanding the role of small enterprise may be a means of increasing the absorption of labor in the course of development. However, neither total employment nor relatively high ratios of labor to output in themselves warrant the conclusion that small firms should be a preferred instrument of development. That judgment depends on all the criteria that are chosen and the degree to which small-scale firms satisfy each of them. The following sections will assess the characteristic organization of small firms, the implications of that organization for the technological choice criteria used, and the potential importance of such firms in the development process.

The service sector is a congeries of firms producing different outputs, with different technologies and different organizations. It includes the small retailer, the great financial institution, the corner shoe-shine boy, the international hotel, the humble carpenter, and the large construction firm. It is not possible in this brief compass to deal with all the issues involved in choosing appropriate technologies for all types of establishments. The present survey is also constrained by the relatively few careful studies of service industries. Special attention will be given to the construction sector because of its particularly critical role in development and because recent investigations permit

TABLE 6 Proportion of Total Employment in Firms Engaging Less than 10 Persons[a]

Sector	Philippines (1961)	Taiwan (1961)	South Korea (1966)	Thailand (1964)	Singapore (1966)
Construction	93	95			
Commerce	94				
Manufacturing	76	46	43	70	45
Transport and Communications	64	58			
Private Services	95	93			

[a] Data taken from Harry T. Oshima, "Labor Force 'Explosion' and the Labor-Intensive Sector in Asian Growth," *Economic Development and Cultural Change*, vol. 19, no. 2 (Chicago: The University of Chicago Press, January 1971), pp. 161–183.

some limited generalizations about the choice of appropriate technologies in some types of construction.

Characteristics of Small-Scale Enterprise and Choice of Appropriate Technology

ORGANIZATION OF SMALL ENTERPRISE

The organization of small enterprise runs the full gamut from the limited liability corporation to "unorganized" household activity with, perhaps, a few nonfamily workers who may or may not be paid regular wages. It is in such enterprise that traditional forms of organization involving various degrees and kinds of partnership and family association are typically found. The Indian study referred to previously shows that almost two-thirds of manufacturing employment is in rural and urban enterprise using mainly household labor. A more recent study also indicates that the proportion of household labor increases among smaller firms.

Even nonfamily labor used in small enterprise may not be compensated on a regular wage basis as, for example, apprentices working in the small establishment. Apprentices, other child labor, and the visiting cousin may be given only food and housing in return for their work. Although the apprentice system may be considered a device by which the "master" pays the apprentice in training instead of wages, there is as yet no evidence that the costs and benefits of the training are effectively balanced by market forces as would be expected in a reasonably well-functioning labor market. Similarly, there is no market test for the supply of capital and other resources to small firms provided directly from family sources.

An example of the difficulties of matching the returns to the various resource inputs used in production in small-scale family enterprise is provided by the treatment of unincorporated enterprise in the national income and output accounts for the United States. Although these aggregate accounts are painstakingly prepared, no attempt is made to distinguish the returns attributable to labor from those attributable to capital and to other resources. Only the total net income of such unincorporated enterprise is estimated, even though in the United States much of it is comparatively large scale and the recordkeeping is relatively good because of the standards imposed by income tax enforcement. Even so, when labor and other resources are provided on a personal and family basis and there are no active market tests as there are when resources are hired and purchased, returns cannot be attributed to the various resources with any degree of confidence.

Another important aspect of the diversity of small enterprise is the ingenuity that some countries and some industries show in achieving within small enterprise the advantages of specialization often attributed to factory work. It is tempting to argue that one of the characteristic differences between "traditional" and "modern" small-scale enterprise is the degree of vertical integration. In one conventional picture of traditional small-scale enterprise, the artisan buys his raw materials—leather, cloth, iron, or wood—and turns out a finished product—shoes, clothes, implements, or furniture. In another conventional image, however, the weaver uses machine-spun thread on his loom and produces a cloth that is then taken to be finished, dyed, and prepared as a final product in other shops.

The putting-out system in which raw and semifinished materials were supplied to small, usually family, workshops on contract was important in the preindustrialization periods in Europe and has its modern counterpart in many countries. Small-scale enterprise has flourished under this system and has expanded from the traditional sectors such as textiles and to the production of the new machine, plastic, and electronic products of the post-World War II period. An important lesson of this recent experience is that this old form of production enterprise can be adapted to modern technological requirements by subdivision of processes and specialization.[1] This adaptation has been achieved in some cases with the assistance of special organizations of producers that have helped to standardize products and set quality and performance standards.

TECHNOLOGICAL CHOICE CRITERIA AND TECHNOLOGICAL DECISIONS IN SMALL-SCALE ENTERPRISE

When the payments to the various resources used in an enterprise are determined to a substantial degree independently of market forces, then to some extent the choice of technologies used can also be independent of market forces.

The objective of small family enterprises using family labor, as noted previously, is not the maximization of profits in the conventional sense of revenues minus costs, including wage costs of all labor. Thus, the profit maximization or cost minimization criterion of appropriateness enforced by competitive markets may not be relevant. Rather, for the family labor that cannot be eliminated for social reasons, the objective is to maximize revenues minus expenditures on hired and purchased inputs. This criterion, as noted in the discussion of traditional agriculture, is likely to lead to relatively intense use of labor. While this has both desirable and undesirable consequences, depending on one's point of view, it means that it is impossible to conclude that the observed use of resources by small enterprise is either physically or economically efficient.

Indeed, some research suggests that small-scale enterprise functions as a reservoir of labor, absorbing it when activity and employment possibilities in the organized sector are reduced and releasing it when the organized sector expands. In some countries, the small enterprise appears to provide a haven for labor released by large enterprise in slack periods.[2] Yet small enterprise can do that only if its labor is casual or family labor and it does not have to meet the same market test for wage payment as does large-scale enterprise.

An alternative interpretation of labor absorption by small-scale industry is that, while small enterprise is fully capitalist in its motivations and use of resources, it faces a different set of labor and capital prices than does large-scale enterprise. First, small enterprise may be less subject to the wage pressures of organized unions; they may also be less effectively subjected to enforcement of minimum wage and other social welfare laws that add to labor costs. Secondly, it is often claimed that small enterprise must pay higher interest rates for loans from organized financial institutions when such credit is available to them at all or must pay higher interest rates in the informal financial markets. These conditions would tend to induce them to use relatively labor-intensive, capital-saving methods when those are available. Unfortunately, in the studies of small enterprise, the importance of these influences has not been distinguished from the significance of its particular structure and family participation, and it would be difficult to do so.

As individual enterprise, small-scale firms have no direct interest in helping to achieve the overall development goals of the economy. So growth maximization, balance of payments correction, regional development, income redistribution, and political development cannot be expected to be goals for small enterprise. If such goals are to be achieved through individual decisions, the government must create incentives through its fiscal or regulatory powers.

A special folklore is that the yeoman peasant family is absorbed and finds satisfaction in a life close to nature and that the sturdy urban artisan and his family likewise mutually find satisfaction and pride in the finished product of their labor. Thus, small enterprise is sometimes associated with achievement of the particular goal of self-sufficiency and a preferred quality of life. Yet there is little organized evidence to indicate whether and how the families choose technologies to advance their particular noneconomic, family goals. Moreover, while families are expected to be at least implicitly concerned with maintaining useful social roles for all its members and sensitive to the quality of life of its members, clearly all family members do not always receive equal consideration in this respect.

Self-sufficiency is not a necessary characteristic of successful small enterprise; some have always been specialized, and in recent years in some countries the success of small enterprise appears to be related to the degree it has specialized. In considering whether families choose technologies to improve the quality of life, it should be recalled that the factory and capitalist enterprise was not the first exploiter of child labor and women's labor; it was the family. One has only to observe children working at looms on the damp floors in the dark back rooms of bazaar shops to understand that the elimination of child labor, the protection of women workers, and improvement in working conditions has been associated with regulated factory enterprise, not small-scale firms.

POTENTIAL FOR LABOR ABSORPTION IN SMALL-SCALE ENTERPRISE

It has been argued that small-scale enterprise provides particularly significant possibilities for development by labor-intensive methods and that it will reduce, if not fully eliminate, the emerging problems of unemployment. The evidence cited consists of the higher labor–output ratios and lower capital–output ratios that often appear to characterize small-scale industry. However, several questions must be answered before these "facts" can be accepted as evidence of the significant potential for labor absorption.

The first question is, "Are the facts correct?" As noted, measuring labor inputs, when family labor is used instead of hired labor paid regular wages, is difficult. The records of labor use in small firms are typically incomplete, if they exist at all. The family labor may not all be working at the intensity of a regular wage recipient. Labor inputs may even be underreported if some of it is supplied by children who should be in school. Table 7 reviews for a number of developing and developed countries the proportions of workers who are employees and presumably paid regular wages and those who are family workers, workers on "own account," and employers for whom records are unreliable. Clearly, the latter percentages are important in these developing nations.

Measuring capital inputs is also difficult. Small enterprise may tend to use more second-hand equipment than larger enterprise, and that equipment is more difficult to value than new equipment. Small enterprise may also use some part of the family home for production or storage. Since "plant" typically constitutes 40 to 60 percent of total capital requirements, underreporting those inputs could well result in capital saving. Perhaps using home buildings for production should be considered a desirable use of excess housing capacity, but this judgment must reflect social as well as private objectives for housing. Casual empiricism suggests that in many developing countries housing conditions in the small-enterprise sector of the economy are seldom desirable. In any case, any substantial expansion of small-scale enterprise in the course of development will require either more "housing" than otherwise envisaged or lower-quality accommodations than planned if part of the housing is used for production facilities.

The next question is, "What are the implications of the information that is available?" As noted earlier, when enterprise is based on the family, or when conditions are noncompetitive, the flourishing of small enterprise cannot be assumed to imply lower real costs of production. The direct evidence on the efficiency of production in small-scale enterprise is at best equivocal. A study for India found that the smallest enterprises tended to be the least efficient.[3] However, generalizations will come slowly at best, and they may never extend beyond the particular country or perhaps the particular sector for which the study is made.

Yet it would be important for policy purposes to know the comparative efficiency of enterprise of various size. If small-scale enterprise is both efficient and relatively labor intensive and capital saving, then promoting it requires no sacrifices of output while providing employment. If that were the case, it would raise the question as to why firms ever become large. On the other hand, if small enterprise is relatively inefficient, it may nonetheless be promoted as a socially acceptable way of absorbing labor; but the costs of using it should be recognized.

Service Sectors, with Special Attention to Construction

In some service activities, the border between employment and unemployment becomes so blurred that the distinction is virtually lost. The men who sell pencils on street corners and the shop girls who run errands or wait on a customer twice a day are only nominally employed. Expanding these activities contributes little, if anything, to development, though they may be acceptable ways for unemployed persons to pass the time and may serve to redistribute income slightly.

However, the use of apparently labor-intensive methods of produc-

TABLE 7 Distribution of the Economically Active Population According to Employment Status: Total and Sectoral (Percent)[a]

Developing Country and Sector	Employees	Unpaid Family Workers	Workers on Own Account	Employers	Not Classifiable
Ecuador					
Manufacturing	39	6	52	3	0
Commerce	21	4	72	3	0
El Salvador					
Manufacturing	60	3	28	4	5
Commerce	38	3	55	3	0
Ghana					
Manufacturing	22	2	76		0
Commerce	12		84		0
Greece					
Manufacturing	64	3	26	6	0
Commerce	35	6	49	10	0

Honduras					
Manufacturing	55	6	37	1	0
Commerce	41	4	53	1	0
Iran					
Manufacturing	64	3	30	2	0
Commerce	21	3	73	2	0
Malaysia					
Manufacturing	66	3	31	2	0
Commerce	41	4	55	2	0
Mexico					
Manufacturing	82	0	17	1	0
Commerce	38	1	59	2	0
Morocco					
Manufacturing	50	3	39	2	5
Commerce	26	3	63	3	5
South Korea					
Manufacturing	64	8	28	2	0
Commerce	13	9	78	3	0

a Data taken from Amartya K. Sen, *Employment, Technology, and Development* (Oxford, England: Oxford University Press, 1975), pp. 18–21.

tion in the service sector has continued to attract attention to the sector, and particularly to the construction industry because of its critical role in the capital-creation process. Substitution of relatively unskilled labor for capital or professional manpower in other service sectors is also of interest. In medical services, for example, attention has been given to methods of substituting technicians for relatively highly trained doctors, mainly to expand medical services rapidly rather than to provide employment.

Output in the construction sector typically runs 3–8 percent of gross national product in developing countries, but its significance for economic growth transcends this percentage. Every investment has a construction component that may be 60 percent or more of the total, and much of the construction must be completed before any equipment can be installed. Thus, to avoid becoming a major bottleneck, the construction sector must expand its output to match the planned expansion of investment on which much of the growth in developing countries depends. Unfortunately, in many cases construction activity has not expanded at the necessary pace and the slow growth of this sector has retarded the development of the entire economy.

In the construction sector, as in agriculture, a wide range of technologies can be used in production. The outputs, if not identical, are quite similar. In some circumstances, labor, using picks and hoes and head baskets, may be substituted for large earth-moving and concrete-mixing and delivery equipment in heavy construction. In addition, it may be feasible to change designs to facilitate use of one or another mix of capital and labor. Earthfill dams built with labor-intensive methods may be substituted in some locations for concrete dams requiring more capital-intensive techniques. Much depends on the specific conditions and requirements placed on the dams or other construction.

Though casual observation shows that there are technically feasible construction techniques with different capital and labor intensities, it is not possible on this basis to assess the degree to which the alternatives are efficient and satisfy the various criteria of appropriateness. Such an assessment requires careful investigation, and only a few studies are available. So generalizations are not yet possible.

Careful research by a group at the World Bank on civil construction methods has provided some interesting results. In addition, it has illustrated the difficulties in assessing the implications of alternative techniques for just the output and cost criteria of appropriateness. The research has demonstrated the importance of soil and terrain conditions in the choice of least-cost technologies, as well as the importance of the project's other particular characteristics. It concluded that the techniques currently in use were either extremely labor intensive or equipment intensive with few intermediate methods available. "Mixed" methods cost more than either. The most labor-intensive methods required more personnel management and supervision and, apparently, longer advance planning. They also typically cost more than the modern capital-intensive technologies, even when labor was priced below market wages in some Asian countries.[4] While these are significant results, this study did not investigate in detail the use of engineering designs that might be better suited to labor-intensive methods. This avenue is currently being explored.

For criteria of appropriateness other than cost, almost no information about the implications of alternative methods is available. While the World Bank study shows that employment can be increased at the expense of higher overall costs in highway construction, the growth and income distribution implications of the alternative technologies were not investigated. As is generally the case, these implications depend on the other policy instruments of the government, and pursuing those issues would take such studies of technology far afield. Analysis of the balance of payments implications of alternative technologies also requires extending the range of investigation to nontechnical issues. For example, while capital-intensive construction methods will often require the import of equipment not produced in most developing countries, labor-intensive methods may require the import of additional consumption goods for the ultimate compensation of labor. The relative weight of one or another import cost will depend on local production and supply. Again, generalizations are difficult.

In relation to political development and national political goals, it is also plausible that alternative technologies should have different effects. Organizing large numbers of workers in labor-intensive methods on construction projects imposes more difficult management problems. In addition, it may impose special demands on political organization, depending on the mixture of economic and political incentives used to mobilize labor. However, highway and other construction also provides opportunities for government to demonstrate its ability both to penetrate the society and to distribute benefits. Yet, while alternative technologies in construction may have political implications, little research has been done on these issues. Thus, it is impossible at this time to go beyond recognizing that such implications may exist.

The most that can be said about the effects of alternative construction technologies on the quality of life is, again, that there may be differential effects. As noted, some observers associate desirable individual and community life-styles with labor intensity and self-sufficiency. In housing construction, designs and production methods have been developed expressly to permit nonspecialized individual and family labor to play a major role in construction. These designs have even been carried over into actual projects, but no overall evaluation appears to have been made. Beyond demonstration of feasibility, evaluation of the trade-offs between self-sufficiency and the other criteria of appropriateness has been relatively limited.

Assessing the potential use of alternative technologies in health delivery and education systems immediately raises problems that are analogous to those in the analysis of construction techniques, but even more intractable. These are the problems of defining the quality of the "outputs" produced by alternative techniques and organizations and the substitutability of different output qualities. Casual observation again reveals alternative technologies that have different relative resource requirements. Modern hospitals with centralized medical services provide one organizational and technological pattern of health care delivery. Neighborhood and rural clinics staffed mainly by technicians with only limited training have different resource requirements. But the alternatives also deliver different outputs. For example, the most decentralized types of health care delivery systems presumably will include some central hospital facilities, and their use would involve higher transport costs than a system with more hospitals. But hospitals, in turn, are more costly than clinics that produce a lower output "quality." Similar comparisons and problems arise in comparing methods in the field of education and in other personal services.

Medical and educational alternatives may have not only different economic costs but also different implications for the distribution of the related components of income that are of particular social and individual importance. In addition, the alternatives may create different demands and opportunities for government and therefore contribute differently to political development. Yet there appears to be little, if any, research that provides direct insights into these important issues.

Another important branch of the service sector provides marketing and storage functions. Since these are often cited as bottlenecks to the improved distribution of agricultural products and achievement of greater benefits from increased agricultural production, they require attention. The range of technological alternatives in these areas and the significance of the alternatives for the achievement of development goals has also not yet been carefully investigated.

Thus, in the construction and service sectors, it seems possible in principle to employ relatively labor-intensive methods when that appears desirable. However, the qualities of the output, the economic efficiencies of the alternative methods, and their contributions to the other criteria of technological appropriateness have not yet been firmly established by careful study.

REFERENCES AND NOTES

1. G. Ranis (1974), pp. 74–75.
2. S. Berger (1974).
3. P. Dhar and H. Lydall (1961), p. 84.
4. C. Harral *et al.* (1974).

8 Policies for Promoting Choice of Appropriate Technologies

Introduction

There can be no doubt that development processes interact intimately with the choice and use of technological methods. However, the preceding discussion also leaves no doubt that understanding of such interactions is at present limited and confined mainly to a narrow economic sphere. In addition, there are a number of alternative criteria of appropriateness for the choice of technologies that are, to some extent, competing rather than complementary. Formulating policies for the choice of appropriate technologies is a particularly difficult task in such circumstances. Nevertheless, the issues are too important to be avoided and, in fact, cannot be, as important technological decisions are being made every day implicitly if not explicitly. Although the basis for making policy is much less adequate than would be desirable, it must be attempted.

The consequences of technological decisions for development, as well as the limited knowledge about them and the potential contradictions in goals, have important implications for the kinds of policies that are desirable. First, the clearest consequences of the particular technologies are their direct effects on output and costs. When there is economic inefficiency there are clear economic losses imposed. Second, it is important to recognize the limitations of technological decisions as policy instruments to satisfy other criteria of appropriateness. While technological factors have many interrelationships with economic and political development, those relationships are often tenuous and indirect. Third, with so much uncertainty and so much risk of error with profound consequences, it is particularly desirable to avoid putting all one's eggs in a single policy basket. Rather, a "mixed portfolio" of policies, accompanied by special efforts to avoid contradictions, is preferable. No single policy emphasis will resolve all the issues in making appropriate technological decisions. In general, policy design should take into account not only the distinctive development goals of each country but also the character and power of the policy instruments that can be used, as well as the special features of each sector. Finally, there is not enough knowledge to prescribe the technological details of processes that would satisfy to a greater degree many of the criteria of appropriateness. In particular, there is little basis for the contention that there is now or, with reasonable assurance, there can be created a set of village-level technologies that can resolve the fundamental development problems. Therefore, at this time, the objectives of policy for appropriate technological decisions must be primarily to increase the knowledge base and to improve the conditions under which the decisions are made to reduce the weight of influences obviously counterproductive to development.

Poverty and high rates of population growth are the fundamental sources of the problems that make the appropriate choice of technology so important. So, the most fundamental policies to deal with those problems are those that, in general, will promote economic development and help to control population growth. Such general policy measures will not be dealt with in this chapter. They are often long run in the sense that their influences on employment and on the other issues of appropriate technology can be expected to manifest themselves only slowly, even when they are effective. However, economic growth and population policies are also short run in the sense that, if they are to have a long run and continuing effect, they cannot be delayed indefinitely.

The policies that will be reviewed below focus specifically on technological choices; they are mainly short run in that they are expected to begin showing results within a few years. They also deal with particular problems that even a rising tide of growth may not wash out.

Three general areas of policy action will be considered. The first is concerned with economic policies that improve the technological selection process by operating on the price, tax, and subsidy incentives affecting the relative use of resources. The second type of policy action concentrates specifically on technological developments that expand the scope of choice in directions suitable to the developing countries. The third policy area is institutional change, which deals with the dissemination of knowledge and the organization of production activities. In fact, of course, successful promotion of appropriate technological choices is often likely to involve effective mixtures of all three types of actions. So the division is somewhat artificial; while the distinctions will facilitate the discussion, they should not be considered mutually exclusive in terms of effective action.

The policy suggestions made are not intended and should not be interpreted as definitive courses of action that will resolve the issues related to appropriate technologies. Rather, they are general policy directives, which, for implementation, must be given specific content for particular countries by more detailed studies. Similarly, the research suggested is intended to exemplify rather than exhaust the topics worthy of investigation.

Economic Policies to Improve Choice of Technologies

Only a few—but potentially quite important—economic policies can be suggested for implementation without further research. First, in evaluating projects, the objectives or criteria for the technological choices in each project should be clearly identified. If the criterion is not to minimize cost, then the efficacy of technological decisions in achieving the other goals should be assessed critically. The relative advantages of using policy instruments other than technological choices to achieve the desired goals should be considered.

If the objective of individual technological decisions is to minimize the costs of the desired outputs, then either the price system must provide the incentives to achieve the objective or government must actively intervene with taxes and subsidies. The rationale is simply that price incentives must be consistent with objectives so that the incentives operate in the desired directions. That is also the rationale for tax and subsidy programs that actively intervene to create "shadow" prices, which provide the desired incentives when actual market prices do not reflect the real relative scarcities in each economy or when the project is explicitly intended to achieve some goal other than economic efficiency. However, as noted earlier, many governments are not able to operate a comprehensive tax and subsidy program with the timeliness and flexibility necessary to achieve desired goals. On the other hand, comprehensive price-control schemes through taxes and subsidies may not be required if only a relatively few projects have to be managed.

Besides enforcing the "right" prices, techniques of cost–benefit analysis must be used for the choice of projects. In economies with relatively limited government supervision of investment decisions, banking and financial institutions should be encouraged to use cost–benefit analysis in their decision making. For projects under government supervision, cost–benefit analysis should be routinely adopted.

In addition to these direct and indirect economic tests to promote the desired technological decisions, consideration should be given to other policy instruments. Project assistance may be made conditional on the provision of evidence that alternative technologies and designs have been evaluated in order to choose the most appropriate one. Investment programs can be examined to determine if they contain unnecessary biases against small-scale enterprise and, again, conditions can be imposed against such biases.

Clearly, a great deal is still unknown about the influences that determine technological choices in developing countries. Though important research has been done and is under way, more should be encouraged. The relatively high cost and long time span required for successful research in this area, as well as its importance, warrants support by the national and international assistance agencies themselves. Examples of important economic research that should be undertaken include:

- Studies of the relations between the institutional means by which information is provided and the exercise of influence over technological decisions;
- Assessment of the explicit and implicit costs of technological information;
- Evaluation of the influence of the structure of private and public management on technological decisions;
- Appraisal of the efficiency of small-scale enterprise in the various sectors.

The list of such relevant topics for economic research is long. Relatively little has been done so far, and the work that remains should not be put off.

Policies for Technological Development

Since in some situations the available techniques appear "inappropriate" by one or more of the standards proposed, suggesting new institutions and facilities to find "more appropriate" technologies may appear natural. However, before taking this step it is first desirable to consider carefully what can reasonably be expected to be accomplished and by what means.

GENERATION OF PRIORITIES FOR THE SEARCH FOR APPROPRIATE TECHNOLOGIES

Consideration should be given to establishing task forces of experienced engineers and economists who would survey production methods and recommend research priorities for particular sectors. They would judge the potential of research activities that would extend the knowledge of efficient technical alternatives in directions leading to more appropriate technologies for development. This needs to be done in each developing country to establish a research agenda for its scarce scientific and engineering manpower. The industrialized countries can both participate in such exercises and undertake independent appraisals. They can also mobilize additional skills in the assessment process and perhaps in the actual research as well, and they may be able to identify some common problems on which cooperative efforts would be warranted.

The diversity of strong views on the possibility of expanding the range of appropriate technology reflects the inadequate state of knowledge in this area. On the one hand, some observers believe that there exists a wide range of alternative, efficient resource combinations from which choices can be made. They conclude, therefore, that major programs of research and development are not required to find more appropriate technologies. In their view, the technologies desired already exist and the problem is to implement them.[1] Moreover, they say, the major research programs envisaged are costly in terms of especially scarce engineering resources that could be better used elsewhere.

On the other hand, some development analysts believe that there is now only a narrow range of efficient alternatives, at least in the major manufacturing sectors;[2] however, even this view does not exclude technological research in order to find new and efficient alternatives. The modern Gandhians, for example, take the position that the current range of technological choices is restricted and must be enlarged; they emphasize the importance of research and development activities to develop intermediate technologies to achieve more intensive use of labor.[3]

The suggestion to look for technologies that are more labor intensive, or generate more savings, or do better in satisfying one or more other criteria of appropriateness is hardly controversial. But, in itself, the suggestion is not helpful either. There are many places to look and not enough resources to look everywhere. Thus, one of the first tasks is to establish priorities for such research. The customary recommendations for technical research are either sweeping and indiscriminate or reflect particular interests and preferences without a general rationale. The proposals for emphasizing self-sufficiency and the preservation of village life, for example, have never presented any estimates of the opportunity costs in giving up the gains from market participation.

Strictly speaking, establishing priorities requires giving relative weights to each of the criteria of appropriateness identified above. This is simply not possible for a research unit because of the essential value judgments involved. But it might be feasible to establish rankings for alternative research programs in terms of potential technical feasibility of creating methods that are improvements in terms of each of the criteria. The process might, at least, indicate where ignorance lies.

Generating such a priority list of technical areas for research would appear to have a good chance of success and would serve several functions. First, it would provide a more tangible basis than now exists for judging the potential returns for research in a new technological direction. Second, it would help provide concrete guidelines for individual researchers and research institutions to orient their research programs. It would also guide assistance agencies in the allocation of their funds for technological research. In the absence of such an overall review of research priorities, an individual project can be evaluated only on its own engineering merits, but no judgment can be made about its relative priority. An economic evaluation is essential in the developing of a priority list as well as in appraising a new technological development.

CREATION OF RESEARCH ORGANIZATIONS

To obtain useful results, a priority list of technological research projects must be followed by actual research. The organization and techniques of such research itself raise many issues beyond the scope of this report.[4] However, a striking difference may be noted between the characteristics of the research now being done on intermediate technologies and the research methods considered characteristic of industrialized countries. The research currently being done on intermediate technologies appears, virtually without exception, to be highly applied and developmental; it is not basic research directed toward fundamentally different approaches to production. There is an assumption in some discussions of intermediate technology that scientifically sophisticated production methods are less "appropriate" than cruder techniques, at least in their employment of labor. While never verified, this implicit assumption may account for the research approach that has typically been adopted.[5] By contrast, the agricultural research in the developing countries has been based on the most advanced scientific methods. The issues involved in the choice of research methods for developing intermediate technologies have hardly been considered yet. This must be done before any research program is committed to a particular approach.

While there are few general guidelines for the organization of new technological research, the history of agricultural research may again provide some useful precedents. One lesson, which is usually emphasized, is that the success of agricultural research depends on its close association with the problems it is expected to resolve. That lesson is also likely to be important for manufacturing and service activities that depend to a great extent on the characteristics of local materials and organizations. Therefore, intermediate technology research in institutions located in industrialized countries may be much less useful than research in national organizations oriented toward the problems of particular industries. On the other hand, for goods and processes whose methods are not location specific, important economies of scale in research may argue for centralization.

The discussion above indicates the many doubts that now exist about the efficacy of major initiatives for research to find new intermediate technologies. Given the risks, it would appear reasonable to consider the research agenda carefully and, at least at the outset, to keep the scale of new research organizations small and closely related to specific tasks.

DISSEMINATION OF INFORMATION ON TECHNOLOGICAL ALTERNATIVES

Another recommendation that has emerged from other considerations of the problems of appropriate technologies is that new institutional channels should be created to disseminate information about existing technologies and the existing range of resource substitution possibilities. This recommendation is based on observation of apparent failures to appreciate the range of choice that already exists in choosing technologies. There are two stages in the successful dissemination of technological information about alternative technologies. The first is to collect or generate useful information; the second is to actually put the information in the hands of potential users.

The collection of technological information on production processes already in use documents the existing range of technological decisions. As noted previously, those decisions may be systematically biased by price distortions, lack of knowledge of alternatives, and irrational bias for sophisticated technologies and other influences. Analogously, concepts or ideas about intermediate or appropriate technologies may be subject to a bias in favor of "simple" techniques. While some of these latter proposals may be eminently meritorious, they also must have their technical and economic feasibility tested.

Unfortunately, the same factors that obstruct the flow of technical information to producers in developing countries tend to create difficulties in the local evaluation of these techniques. There may be only limited national expertise for technical and economic analysis of alternative techniques before they are actually adopted and implemented. The advice of foreign experts, however, may reflect their implicit or explicit preferences for methods familiar to them.

Thus, the mere dissemination of information about the existence of alternative technologies, either actually in use or only in a conceptual stage, may have only a limited utility, if it is not, in fact, inadvertently misleading. To avoid the latter possibility, all the reservations necessarily associated with the technical information disseminated should be emphasized.

The dissemination of information about technological alternatives would be more useful if it were accompanied by technical and economic evaluations. Since such evaluations can be a major task of engineering and economic analysis, consideration should be given to making this one of the major undertakings of the research organizations discussed previously.

Dissemination of technological information itself can proceed at several stages. Through the studies and conferences it sponsors, the United Nations Industrial Development Organization generates a great deal of technological information that is made widely available. However, only rarely could an individual manager or engineer in a developing country be expected to be aware of that information and to take the trouble to locate and use the proper data. For such collections of information to be useful, an effective set of institutions is necessary to bring potential users into contact with it. The problems involved in achieving intensive use of relevant technological information will be discussed below in connection with institutional changes to achieve more appropriate use of technologies.

The major obstacle to recommending specific new programs and institutions for information dissemination is the limited understanding of the effectiveness of the many public and private channels that now exist. Therefore, the formulation of public policies to improve the collection and dissemination of technological information should start with an assessment of the effectiveness of current efforts. That requires the identification of active institutions, the functions and economic sectors in which they specialize, and their mode of operation.

Policy formulation also requires assessment of the relative effectiveness of the various types of private institutions that disseminate technological information: equipment manufacturers; construction firms; and engineering, economic, and management consulting firms. Although a great deal of experience has accumulated in the developing areas, it has seldom been evaluated. Indeed, there is no evidence that the various types of public and private institutions have performed ated their own operations thoroughly and in depth. These agencies should consider launching research programs to determine how well the various types of public and private institutions have performed separately and together in the investment programs that have been sponsored.

The identification and engineering evaluation of information collected and disseminated on various production processes is an essential technological component of the recommended research. However, the economic and organizational aspects of research on the effectiveness of information dissemination methods are equally important. Thus, a balanced approach, including social scientists as well as engineers, is again necessary.

Institutional and Political Policies for Choice of Appropriate Technologies

While the choice of technologies to achieve particular political goals and the political impacts of technological choices have been discussed, so little is known of the relations involved that direct policy recommendations in this area are not now possible. Therefore, it is again necessary to emphasize new research. The political consequences of technological decisions may have direct effects on the organization and use of political influence. However, the most important political consequences may be indirect and operate via urbanization, general economic growth, and income distribution. These issues require much more study.

Once the importance of political goals in development is recognized, then it becomes clear that the choice of economically inappropriate technologies may not be simply the result of ignorance about the consequences of the choice. Rather, it may reflect the dominance of political considerations, including the possibility that decision makers were unable to resolve conflicting objectives in any other manner than through economically inappropriate decisions.

While there is only limited general understanding of the relations between technological decisions and political issues, the political implications of particular projects may be clearly perceived—although it may be inconvenient to recognize them publicly. This may be as true for national and international assistance agencies as it is for the governments of the developing countries. Where such political factors operate, it would be naïve to call for the dominance of economic considerations. However, it is desirable to consider whether an alternative policy that did not involve the distortion of technological choices would be feasible and effective. Research would be desirable on the conditions in which political considerations dominate technological decisions.

The significance of the institutional structure of production organizations in determining the way resources are used is another relatively neglected area of empirical research. The most intensively studied institutions are profit-maximizing, capitalist firms. Theoretical economic studies of alternative organizational forms are, however, becoming more frequent. These include studies of sharecropping and other forms of tenancy and worker-managed firms. Such research should be supplemented with political and sociological studies, including empirical investigation, since the economic rationale of the particular institution both depends on, and contributes to, political and social structures and functions. For example, it has been claimed that trade associations have been important in determining the ability of small enterprise to survive and prosper and expand into the production of modern producer and consumer goods. But this requires more study. Further investigation is also required to assess the practical significance of government standards and services in determining the survival of small enterprise.

Investigation of Potential for Improvements in Licensing Procedures

Negotiations for technology transfer are currently almost exclusively bilateral. Each of the receiving countries, represented either by government or private organizations, in turn deals with other government or private organizations. Typically, the receiving countries have only limited information about the relative advantages of alternative technologies available for particular products and about the costs of acquiring the rights to use the technologies. The owners and sellers of technology, on the other hand, are usually relatively well informed. While knowledge about the most basic features of technology is relatively widely available, the detailed information embodied in special types of know-how and even in particular pieces of equipment is often closely held. In other types of production, the basic patents and necessary equipment can be obtained from only a few sources, often just one.

While effective competition among technology sellers would—by definition—eliminate any advantages the sellers might have, there is little to suggest such competition actually exists and, on *a priori* grounds, it would not be expected. The sellers' advantages may, in part, explain why the fee schedules for technology purchases do not appear to contain provisions that allow the buying countries to recapture the benefits that their experience contributes to the sellers.[5] The sellers of technology emphasize the costs of the research and the importance of royalty and licensing fees to stimulate additional research. When royalties or licensing fee payments depend on the rate of production or income, the returns to technology sellers are also subject to the uncertainties associated with new enterprises or types of production.

There are clearly inequalities in bargaining power among the buyers and sellers of technology. There are also potential inefficiencies when transactions for the same technological information are carried out on a project-by-project basis rather than on a basis that recognizes a wider demand for particular technologies. The limited information about contracts for technology puts the buyer at a disadvantage, but in some circumstances it may also operate to the disadvantage of the seller.

Although the problems and imperfections in the markets for technological information are apparent, the best means for their resolution are not clear. The existing modes of transactions, by individuals and/or by countries, are not optimal. If a new international agency were to assume any of the functions of the present contractors, major difficulties would arise in its establishment and functioning, given the highly particularized character of many of the negotiations and the interests involved. Thus, this is another area where further detailed research is needed before any recommendations can be made.

It is tempting to isolate a particular problem in the development process, such as unemployment, and to attempt to deal directly with that problem by modifying technologies to increase the absorption of labor. However, although the particular problem is a recognizably distinct feature, it is intimately related to many other essential conditions of the economy. Moreover, modifying technologies toward more highly labor-intensive methods may be impossible as well as unnecessary. In some important cases, at least, opportunities already exist for efficient but more intensive use of labor. Thus, while no obstacles should be created to the search for new and better technologies, the highest priorities should be given to policies that would stimulate the pace of development, reduce the pressures of population growth, and improve the dissemination of information about existing technologies and the process of decision making to help each country achieve its particular goals.

REFERENCES AND NOTES

1. As proponents of this view see, for example, A. Sen (1975), also C. Cooper *et al. in* A. Bhalla (1975), pp. 115–116.
2. See A. Kelley *et al.* (1972), pp. 25–26.
3. E. Schumacher (1973), *passim.*
4. See, for example, any issue of *Intermediate Technology.*
5. United Nations Conference on Trade and Development (1975), pp. 57–59.

Appendix

Comments by Dr. Simón Teitel Concerning the Report "Appropriate Technologies for Developing Countries"

While I must side with Professor Richard Eckaus on the matter of "intermediate" (between what?) technology, as well as the inconclusiveness of much of the research done so far on this topic and on the question of technological determinism and quality of life, I must take exception to his basic approach. My main criticism is with the conceptual framework about technology used in the report. While there is some awareness in the paper about the limitations of received economic theory, the method used is still essentially reductive, and technology is for the most part only conceived as *technique* (i.e., factor proportions) and as a datum, a given to the economic system. This, as Stigler stated, assumes away the problem of choice of technology (see G. J. Stigler, "The Xistence of X-Efficiency," *American Economic Review,* March 1976). As a consequence, we have a statement which deals mostly with the choice of sector and product rather than technology. In fact, it has already been obvious from previous work in this field that we need to treat the subject of appropriate technology from the point of view of restrictions to the flow of information, considering both international and domestic institutional constraints (see K. Arrow, "Limited Knowledge and Economic Analysis," *American Economic Review,* March 1974). Instead, the standard two-factor neoclassical model is followed, and as a result no mention is made of such crucial factors as materials, scale, and inappropriate machinery. Although there is some discussion about the transfer of information of a technical nature, the analysis is limited essentially to the flow of information from outside the LDCs, and is lacking in regard to the economics of information. Little is said also about the adaptation and development of technologies in LDCs.

A further problem I have with the paper, also conceptual, is with the meaning ascribed to "appropriate." There are here two basic possibilities, both relevant: "appropriate" to objectives and "appropriate" to the environment. Eckaus deals only with possible criteria for appropriateness from the point of view of alternative growth objectives and does not consider the adequacy of the technology to specific factor endowments, informational restrictions, and other institutional aspects of LDCs. This constitutes not only a problem in itself but is in part responsible for the lack of sufficient emphasis on the extent of adaptation of technology going on in LDCs.

With reference to objectives, the author ignores fundamental political objectives in relation to technology, such as self-reliance and defense, discussing instead some that are of little real relevance to LDCs.

Bibliography

Acharya, Shankar N. *Fiscal/Financial Intervention, Factor Prices and Factor Proportions: A Review of Issues.* IBRD Bank Staff Working Paper no. 183. Washington, D.C.: International Bank for Reconstruction and Development, 1974.

Achilladelis, Basil. *Emerging Changes in the Petrochemical Industry: An Overview.* OECD Development Centre, Occasional Paper no. 3. Paris: Organization for Economic Cooperation and Development, 1974.

Adelman, Irma. "Development Economics—A Reassessment of Goals." *American Economic Review*, Papers and Proceedings, 65 (May 1975):302–309.

Adelman, Irma, and Morris, Cynthia T. *Economic Growth and Equity in Developing Countries.* Palo Alto, Calif.: Stanford University Press, 1973.

Allal, Moise. *Selection of Road Projects and the Identification of the Appropriate Road Construction Technology: General Considerations.* World Employment Programme Research, WEP—22/WP14. Geneva: International Labour Office, 1975.

Appropriate Technology, London, 1 (Autumn 1974).

Arndt, Thomas M., and Ruttan, Vernon W. *Resource Allocation and Productivity in National and International Agricultural Research.* Research and Training Network Seminar Report Series, September 1975. New York: Agricultural Development Council, Inc., 1975.

Arrow, Kenneth J. "The Economic Implications of Learning by Doing." *Review of Economic Studies* 29 (June 1962):155–173.

Asher, R. E.; Hagen, E. E.; Hirschman, A. O.; Colm, G.; Geiger, T.; Mosher, A. T.; Eckaus, R. S.; Bowman, M. J.; Anderson, C. A.; and Wriggins, H. *Development of the Emerging Countries: An Agenda for Research.* Washington, D.C.: The Brookings Institution, 1972.

Aubrey, Henry C. "Small Industry in Economic Development." *Social Research* 18 (September 1951):269–312.

Baeumer, Ludwig. *Importance of Industrial Property Protection in Developing Countries* (ID/WG.130/4). Vienna: United Nations Industrial Development Organization, 1972.

127

Bardhan, Pranab K. "Agricultural Development and Land Tenancy in a Peasant Economy: A Theoretical and Empirical Analysis." Ottawa, Canada: International Development Research Centre, Income Distribution Division, 1975.

Bardhan, P. K., and Srinivasan, T. N. "Cropsharing Tenancy in Agriculture." *American Economic Review* 61 (March 1971):48–64.

Bell, Clive. "The Acquisition of Agricultural Technology: Its Determinants and Effects." *The Journal of Development Studies* 9 (October 1972):123–160.

Berger, Suzanne. "The Uses of the Traditional Sector: Why the Declining Classes Survive." In *Il Caso Italiano,* by Fabio Luca Cavazza and Stephen R. Graubard. Milan, Italy: Garzanti, 1974.

Berry, A. *Unemployment as a Social Problem in Urban Colombia: Some Preliminary Hypotheses and Interpretation.* Yale University, Economic Growth Center, Discussion Paper no. 145. New Haven, Conn.: Yale University Press, 1972.

———. *Urban Labour Surplus and the Commerce Sector: Colombia.* Yale University, Economic Growth Center, Discussion Paper no. 178. New Haven, Conn.: Yale University Press, 1973.

Bhagwati, Jagdish N. *Amount and Sharing of Aid.* Washington, D.C.: Overseas Development Council, 1970.

———. *The Economics of Underdeveloped Countries.* New York: McGraw-Hill, 1970.

Bhagwati, Jagdish N., and Chakravarty, S. "Surveys of National Economic Policy Issues and Policy Research." *The American Economic Review,* Supplement, 59 (September 1969), 118 pp.

Bhalla, A. S., ed. *Technology and Employment in Industry.* Geneva: International Labour Office, 1975.

Binder, L. "The Crises of Political Development." In *Crises and Sequences in Political Development,* by L. Binder *et al.,* pp. 3–72. Princeton, N.J.: Princeton University Press, 1971.

Binder, L.; Coleman, J. S.; La Palombara, J.; Pye, L. W.; Verba, S.; and Weiner, M. *Crises and Sequences in Political Development.* Princeton, N.J.: Princeton University Press, 1971.

Blitzer, C. R.; Clark, P. B.; and Taylor, L. *Economy-Wide Models and Development Planning.* Oxford, England: Oxford University Press, 1975.

Boon, Gerard K. "Economic Technological Behavior in Development." Guanajuato, Mexico: El Colegio de Mexico, 1973.

Brewer, Garry D., and Brunner, Ronald D., ed. *Political Development and Change.* New York: The Free Press, 1975.

Browne, Malcolm W. "Expert Says World Has 27 Days' Food." *The New Times,* August 21, 1974.

Burke, Fred G. *Tanganyika: Preplanning.* Syracuse, N.Y.: Syracuse University Press, 1965.

Cassen, Robert H. "Population Growth and Public Expenditure in Developing Countries." In *International Population Conference Liege 1973.* Vol. 1, pp. 333–346. Liege, Belgium: International Union for the Scientific Study of Population, 1973.

Chenery, H. B., and Bruno, M. "Development Alternatives in An Open Economy." *Economic Journal* 72 (March 1962):79–103.

Chenery, Hollis; Ahluwalia, Montek S.; Bell, C. L. G.; Duloy, John H.; and Jolly, Richard. *Redistribution with Growth.* Oxford, England: Oxford University Press, 1974.

Chenery, Hollis, and Syrquin, Moises. *Patterns of Development 1950–1970.* London: Oxford University Press, for the World Bank, 1975.

Choi, Harry Y. H. *The RANN Program: Potential Benefits to Developing Countries.* Office of Science and Technology Series, TA/OST 73-16. Washington, D.C.: U.S. Agency for International Development, Office of Science and Technology, 1973.

Choucri, Nazli. "Technological Choice and Political Development." Cambridge, Mass.: Massachusetts Institute of Technology, Department of Political Science, 1976.

Coale, Ansley J. "Too Many People?" *Challenge* 17 (September/October 1974):29.

Cohen, Allan R. *Tradition, Change and Conflict in Indian Family Business.* The Hague, The Netherlands: Mouton Publishers, 1974.

Commission on International Development. *Partners in Development.* New York: Praeger Publishers, Inc., 1969.

Cooper, Charles. "Science, Technology and Production in the Underdeveloped Countries: An Introduction." *The Journal of Development Studies* 9 (October 1972):1–18.

Courtney, William H., and Leipziger, Danny M. *Multinational Corporations in LDCs: The Choice of Technology.* U.S. Agency for International Development, Bureau for Program and Policy Coordination, Discussion Paper, no. 29, December 1974. Washington, D.C.: U.S. Agency for International Development, 1974.

Crane, Diana. *An Inter-Organizational Approach to the Development of Indigenous Technological Capabilities: Some Reflections on the Literature.* Organization for Economic Cooperation and Development, Development Centre, Occasional Paper, no. 3. Paris: Organization for Economic Cooperation and Development, 1974.

Dahl, Norman C. "Absorption of Technology in Developing Countries." Paper presented at Symposium on Technology, Modernization and Cultural Impact, May 8–9, 1974, at Iowa State University.

Dasgupta, Asim K. "Two Essays on Income Distribution in a Developing Economy." Ph.D. dissertation, Massachusetts Institute of Technology, 1975.

Dasgupta, Partha; Sen, Amartya; and Marglin, Stephen. *Guidelines for Project Evaluation.* Project Formulation and Evaluation Series, no. 2 (ID/SER.H/2). Vienna: United Nations Industrial Development Organization, 1972.

Denison, Edward T. *The Sources of Economic Growth in the United States.* New York: Committee for Economic Development, 1962.

———. *Why Growth Rates Differ.* Washington, D.C.: The Brookings Institution, 1967.

Dhar, P. N., and Lydall, H. F. *The Role of Small Enterprises in Indian Economic Development.* New York: Asia Publishing House, 1961.

Diamond, Peter A., and Mirrlees, James A. "Optimal Taxation and Public Production." *American Economic Review* LXI (March 1971):261–78.

Diwan, Romesh K., and Gujarati, Damodar N. "Employment and Productivity in Indian Industries—Some Questions of Theory and Policy." *Artha Vijnana: Journal of the Gokhale Institute of Politics and Economics,* India, 10 (March 1968):29–67.

East–West Center. Technology and Development Institute. *Pilot Study on the Generation and Diffusion of Adaptive Technology in Indonesia: A Summary Report.* Honolulu, Hawaii: East–West Center. Technology and Development Institute, 1972.

Eckaus, Richard S. "The Factor Proportions Problem in Underdeveloped Areas." *American Economic Review* 45 (September 1955):539–565.

———. "Education and Economic Growth." In *Economics of Higher Education,* edited by S. J. Mushkin, pp. 102–128. Washington, D.C.: U.S. Department of Health, Education and Welfare, 1962.

Eckaus, Richard S., and Parikh, Kirit S. *Planning for Growth: Multisectoral, Intertemporal Models Applied to India.* Cambridge, Mass.: The M.I.T. Press, 1968.

Edwards, Edgar O. *Employment in Developing Nations.* New York: Columbia University Press, 1974.

Elmandjra, M. *UNESCO Information Exchange and Development: Background Paper for DEVSIS Feasibility Study for the Preliminary Design for an International Information System for the Development Sciences* (DEVSIS/FS/ME/5), March 3, 1975. Paris: United Nations Educational, Scientific, and Cultural Organization, 1975.

Evenson, Robert E., and Kislev, Yoav. *Agricultural Research and Productivity.* New Haven, Conn.: Yale University Press, 1975.

Falcon, Walter. "The Green Revolution: Generations of Problems." *The American Journal of Agricultural Economics* 52 (December 1970):698–712.

Fei, J. C. H., and Ranis, Gustav. *A Model of Growth and Employment in the Open Dualistic Economy: The Cases of Korea and Taiwan.* SEADAG papers on problems of development in Southeast Asia, no. 73-2. New York: Southeast Asia Development Advisory Group of the Asia Society, 1973.

———. *Growth and Employment in South Korea and Taiwan.* New Haven, Conn.: Yale University Press, 1974.

Field, John Osgood. "Politicization and System Support in India: The Role of Partisanship." Cambridge, Mass.: Massachusetts Institute of Technology, Center for International Studies, 1974.

Figlewski, Stephen. "The Burden of Capital Intensive Technology in a Capital-Poor Country." Cambridge, Mass.: Massachusetts Institute of Technology Press, 1975.

The Ford Foundation Letter, 6 (April 1, 1975).

Gemmill, Gordon, and Eicher, Carl. *The Economics of Farm Mechanization and Processing in Developing Countries: Report on an ADC/RTN Seminar, Held at Michigan State University March 23–24, 1973.* Research and Training Network Seminar Report Series, no. 4. New York: The Agricultural Development Council, Inc., 1973.

German Foundation for Developing Countries. *Development and Dissemination of Appropriate Technologies in Rural Areas.* Berlin: German Foundation for Developing Countries, Centre for Economic and Social Development, 1972.

Germidis, Dimitri, and Brochet, Christine. *Le Prix du Transfert de Technologies dans les Pays en Voie de Développement.* Organization for Economic Cooperation and Development, Development Centre, Etude Special, no. 5. Paris: Organization for Economic Cooperation and Development, 1975.

Giral, José Barnes. "Appropriate Chemical Technologies for Developing Economies: A Report Prepared for a Study of Appropriate Technologies in Developing Economies of the Board on Science and Technology for International Development (BOSTID), National Academy of Sciences (NAS) and for the Office of the Foreign Secretary, National Academy of Engineering (NAE)." Mexico: Universidad Nacional Autónoma de Mexico, 1973.

Gotsch, Carl. "Technical Change and the Distribution of Income in Rural Areas." *American Journal of Agricultural Economics* 54 (May 1972):326–341.

Gotsch, Carl; Ahmed, Bashir; Naseem, Mohammad; Falcon, Walter P.; and Yusuf, Shahid. "Linear Programming and Agricultural Policy: Micro Studies in the Pakistan Punjab." *Food Research Institutes Studies* 14 (1975); 104 pp.

Gotsch, Carl, and Falcon, Walter P. "Technology, Institutions and Rural Development: Notes for Discussion." Cambridge, Mass.: Harvard University, Development Advisory Service, 1972.

Griffin, Keith. *The Political Economy of Agrarian Change: An Essay on the Green Revolution.* Cambridge, Mass.: Harvard University Press, 1974.

Griliches, Zvi. "Research Expenditures, Education and the Aggregate Agricultural Production Function." *American Economic Review* 54 (December 1964):961–975.

Griliches, Zvi, and Jorgenson, Dale. "The Explanation of Productivity Change." *Review of Economic Studies* 34 (July 1967):249–283.

Hagen, Everett E. *On the Theory of Social Change.* Homewood, Ill.: The Dorsey Press, Inc., 1962.

Hahn, Albert V. G. *Towards a Reappraisal of the Petrochemical Industry: Technology and Economics.* Organization for Economic Cooperation and Development, Development Centre, Occasional Paper, no. 2. Paris: Organization for Economic Cooperation and Development, 1974.

Harral, Clell G., et al. *Study of the Substitution of Labor and Equipment in Civil Construction: Phase II Final Report.* 3 vols. Washington, D.C.: International Bank for Reconstruction and Development, 1974.

Harris, J., and Todaro, M. "Migration, Unemployment and Development." *American Economic Review* 60 (March 1970):126–142.

Hayami, Yujiro, and Ruttan, Vernon W. *Agricultural Development: An International Perspective.* Baltimore, Md.: The Johns Hopkins Press, 1971.

Hayter, Teresa. *Aid as Imperialism.* Middlesex, England: Penguin Books, 1971.

———. *French Aid.* London: Overseas Development Institute, Ltd., 1966.

Hibbs, Douglas A. *Mass Political Violence: A Cross-National Causal Analysis.* New York: John Wiley & Sons, 1973.

Hoos, Ida R. "Systems Analysis as Technology Transfer." *Journal of Dynamic Systems, Measurement, and Control* 96 (March 1974):1–5.

Horie, Yasugo. "Modern Entrepreneurship in Meiji, Japan." In *The State and Economic Enterprise in Japan,* edited by W. W. Lockwood, pp. 183–208. Princeton, N.J.: Princeton University Press, 1965.

Hughes, Helen. "The Scope for Labor Capital Substitution in the Developing Economies of Southeast and East Asia." In *Technology, Employment, and Development: Selected Papers Presented at Two Conferences Sponsored by the Council for Asian Manpower Studies,* edited by Lawrence J. White, pp. 30–57. Quezon City, Philippines: Council for Asian Manpower Studies, 1974.

Huntington, Samuel P. *Political Order in Changing Societies.* New Haven, Conn.: Yale University Press, 1968.

Inkeles, Alex, and Smith, David H. *Becoming Modern. Individual Change in Six Developing Countries.* Cambridge, Mass.: Harvard University Press, 1974.

Islam, Nurul, ed. *Agricultural Policy in Developing Countries.* New York: Halsted Press, 1974.

Kelley, Alan C.; Williamson, Jeffrey; and Chatham, Russell J. *Dualistic Economic Development.* Chicago: University of Chicago Press, 1972.

Kennedy, Charles. "Induced Bias in Innovation." *Economic Journal* 74 (September 1964):541–547.

Kilby, Peter. "Farm and Factory: A Comparison of the Skill Requirements for the Transfer of Technology." *The Journal of Development Studies* 9 (October 1972):63–70.

King, Timothy. *Population Policies and Economic Development: A World Bank Staff Report.* Baltimore, Md.: Published for the World Bank by the Johns Hopkins Press, 1974.

Korea Institute of Science and Technology. *Final Report on a Study of the Scope for Capital-Labor Substitution in the Mechanical Engineering Sector.* Seoul: Korea Institute of Science and Technology, 1973.

Krishnamurty, J. *Indirect Employment Effects of Investment in Industry.* World Employment Programme Research, WEP—22/WP6. Geneva: International Labour Office, 1974.

La Palombara, Joseph. "Penetration: A Crisis of Governmental Capacity." In *Crises and Sequences in Political Development,* by Leonard Binder et al., pp. 205–282.

Princeton, N.J.: Princeton University Press, 1971.

Leff, Nathaniel H. *The Brazilian Capital Goods Industry, 1929–1964.* Cambridge, Mass.: Harvard University Press, 1968.

Leibinstein, Harvey. "Allocative Efficiency vs. X-Efficiency." *American Economic Review* 56 (June 1966):392–415.

Lele, Uma. "Technology and Rural Development in Africa." Paper presented at the Annual Meeting of the American Association for the Advancement of Science, February 24–March 1, 1974, San Francisco, California.

Leontief, Wassily, ed. *Studies in the Structure of the American Economy.* New York: Oxford University Press, 1953.

Lewis, John P. *Quiet Crisis in India.* Washington, D.C.: The Brookings Institution, 1962.

Light, David. "Elasticities of Substitution and Labor Absorption in the Indian Industrial Sector." Cambridge, Mass.: Massachusetts Institute of Technology, 1974.

Little, Ian M. D., and Mirrlees, James. *Manual of Industrial Projects Analysis in Developing Countries.* Social Cost-Benefit Analysis, Vol. II. Paris: Organization for Economic Cooperation and Development, Development Centre, 1969.

Little, Ian M. D.; Scitovsky, Tibor; and Scott, Maurice. *Industry and Trade in Some Developing Countries: A Comparative Study.* Oxford, England: Oxford University Press for the Organization for Economic Cooperation and Development, 1970.

Machlup, Fritz. *The Production and Distribution of Knowledge in the United States.* Princeton, N.J.: Princeton University Press, 1967.

Macpherson, George, and Jackson, Dudley. "Village Technology for Rural Development: Agricultural Innovation in Tanzania." *International Labour Review* 3 (February 1975):105–126.

Maddison, Angus. *Economic Progress and Policy in Developing Countries.* London: George Allen and Unwin, 1970.

Manser, W. A. P. "The Financial Role of Multinational Enterprises." In *Multinational Enterprises,* edited by J. S. G. Wilson and C. F. Scheffer. Leiden, The Netherlands: A. W. Sijthoff, 1974.

Martin, Edwin M. *Development Assistance: Efforts and Policies of the Members of the Development Assistance Committee.* Organization for Economic Cooperation and Development Review Series, 1971. Paris: Organization for Economic Cooperation and Development, 1971.

———. *Development Cooperation: Efforts and Policies of the Members of the Development Assistance Committee.* Organization for Economic Cooperation and Development Review Series, 1972. Paris: Organization for Economic Cooperation and Development, 1972.

Mason, Edward S., and Asher, Robert E. *The World Bank Since Bretton Woods.* Washington, D.C.: The Brookings Institution, 1973.

Meesook, Kanitta M. "Studies in the Theory of Wealth and Income Distribution." Ph.D. dissertation, Massachusetts Institute of Technology, 1974.

Mendis, D. L. O. *Planning the Industrial Revolution in Sri Lanka.* Organization for Economic Cooperation and Development, Development Centre, Occasional Paper, no. 4. Paris: Organization for Economic Cooperation and Development, 1975.

Michie, Barry H. "Variations in Economic Behaviour and the Green Revolution: An Anthropological Perspective." *Economic and Political Weekly;* a Sameeksha Trust Publication, India, VIII (June 30, 1973):A-67–A-75.

Millikan, Max F., and Blackmer, Donald L. M., eds. *The Emerging Nations.* Boston, Mass.: Little, Brown and Company, 1961.

Millikan, Max F., and Hapgood, David. *No Easy Harvest.* Boston, Mass.: Little, Brown and Company, 1967.

Mishan, E. J. *Technology and Growth: The Price We Pay.* New York: Praeger Publishers, Inc., 1970.

Mishikawa, Shunsaku. "Unemployment Statistics Compilation Needs Revision." *The Japan Economic Journal/Nihon Keizei Shimbun,* Tokyo, no. 13 (May 6, 1975), p. 3.

"MOL Says Labor Market Struck Bottom in April." *The Japan Economic Journal/ Nihon Keizei Shimbun,,* Tokyo, June 3, 1975.

Montgomery, John D. *Allocation of Authority in Land Reform Programs: A Comparative Study of Administrative Processes and Outputs.* Research and Training Network Reprint Series, March 1974. New York: Agricultural Development Council, Inc., 1974.

————. *Technology and Civic Life: Making and Implementing Development Decisions.* Cambridge, Mass.: Massachusetts Institute of Technology Press, 1974.

Morawetz, David. "Import Substitution, Employment and Foreign Exchange in Colombia: No Cheers for Petrochemicals." In *The Choice of Technology in Developing Countries: Some Cautionary Tales,* by C. Peter Timmer, John Woodward Thomas, Louis T. Wells, Jr., and David Morawetz, pp. 95–105. Cambridge, Mass.: Harvard University, Center for International Affairs, 1975.

Morley, Samuel A., and Smith, Gordon W. "The Choice of Technology: Multinational Firms in Brazil." Rice University Working Paper, no. 58. Houston, Texas: Rice University, 1974.

Mowlana, Hamid. "The Multinational Corporation and the Diffusion of Technology." In *The New Sovereigns,* by Abdul A. Said and L. R. Simmons. Englewood Cliffs, N.J.: Prentice Hall, 1975.

Nasbeth, L., and Ray, G. F. *The Diffusion of New Industrial Processes.* Cambridge, England: Cambridge University Press, 1974.

National Academy of Sciences, Office of the Foreign Secretary. *Rapid Population Growth: Consequences and Policy Implications.* 2 vols. Baltimore, Md.: The Johns Hopkins Press, 1971.

National Bureau of Economic Research. *The Rate and Direction of Inventive Activity.* Princeton, N.J.: Princeton University Press, 1962.

Negandhi, Anant R., and Prasad, S. Benjamin. *The Frightening Angels: A Study of U.S. Multinationals in Developing Nations.* Kent, Ohio: The Kent State University Press, 1975.

Nelson, Richard R. "Less Developed Countries—Technology Transfer and Adaptation: The Role of the Indigenous Science Community." *Economic Development and Cultural Change,* 23 (October 1974):61–77.

Nelson, Richard R., and Norman, Victor. *Technological Change and Factor Mix over the Product Cycle: A Model of Dynamic Comparative Advantage.* Yale University, Economic Growth Center, Discussion Paper, no. 186. New Haven, Conn.: Yale University, 1973.

Nerlove, Marc. *Economic Growth and Population: Perspectives of the "New Home Economics."* Agricultural Development Council Reprint Series, November 1974. New York: Agricultural Development Council, 1974.

Norman, Victor D. "Education, Learning and Productivity." Ph.D. dissertation, Massachusetts Institute of Technology, 1972.

Nurkse, Ragnar. *Problems of Capital Formation of Underdeveloped Countries.* Oxford, England: Blackwell, 1953.

Okabe, Naoaki. "Present Unemployment Can Become Very Serious." *The Japan Economic Journal/Nihon Keizai Shimbun,* Tokyo, no. 13, February 25, 1975, p. 1.

Organization for Economic Cooperation and Development. *Choice and Adaptation of Technology in Developing Countries: An Overview of Major Policy Issues.* Paris: Organization for Economic Cooperation and Development, 1974.

————. *Transfer of Technology for Small Industries.* Paris: Organization for Economic Cooperation and Development, 1974.

Organization for Economic Cooperation and Development. Development Centre. *Low-Cost Technology—An Inquiry Into Outstanding Policy Issues: Interim Report of the Study Sessions Held in Paris, 17–20 September, 1974.* Paris: Organization for Economic Cooperation and Development, Development Centre, 1975.

————. *Research and Related Activities of the Development Centre in 1975.* Paris: Organization for Economic Cooperation and Development, Development Centre, 1975.

Oshima, Harry T. "Labour Force Explosion and the Labour Intensive Sector in Asian Growth." *Economic Development and Cultural Change* 19 (January 1971):161–183.

Pearson, Lester B. *The Crisis of Development.* New York: Praeger Publishers, Inc., 1970.

Plasschaert, S. "Multinational Companies and International Capital Markets." In *Multinational Enterprises,* edited by F. S. G. Wilson and C. T. Scheffer. Leiden, The Netherlands: A. S. Sijthoff, 1974.

Poblete, Juan Antonio, and Harboe, Ricardo. *A Case on Transfer of Knowledge, in Water Resources Systems Planning, from a Developed Region to a Developing One and from Research to Application.* Santiago, Chile: Universidad de Chile, 1972.

Prebisch, Raúl. "Commercial Policy in the Underdeveloped Countries." *American Economic Review,* Papers and Proceedings, 49 (May 1959):251–273.

Ramos, Joseph. *An Heterodoxical Interpretation of the Employment Problem in Latin America.* Santiago, Chile: Programa Regional del Empleo Para America Latina y el Caribe, 1973.

Ranis, Gustav. "Some Observations on the Economic Framework for Optimum LDC Utilization of Technology." In *Technology, Employment, and Development: Selected Papers Presented at Two Conferences Sponsored by the Council for Asian Manpower Studies,* edited by Lawrence J. White, pp. 58–96. Quezon City, Philippines: Council for Asian Manpower Studies, 1974.

"Red Light Comes on in Consumer Price Sector." *The Japan Economic Journal/Nihon Keizai Shimbun,* Tokyo, no. 13, June 3, 1975, p. 3.

Rensberger, Boyce. "New Corn Type Exceeds Beef in Quality of Protein." *The New York Times,* September 4, 1974.

————. "Science Gives New Life to the Green Revolution." *The New York Times,* September 3, 1975.

Rice, E. B. *Extension in the Andes.* Cambridge, Mass.: The M.I.T. Press, 1971.

Roberts, John. "Engineering Consultancy, Industrialization and Development." *The Journal of Development Studies* 9 (October 1972):39–62.

Rosenstein-Rodan, Paul N. "Problems of the Industrialization of Eastern and South-Eastern Europe." *Economic Journal* 53 (June–September 1943):202–211.

————. "Rapporti fra Factori Produttivi nell' Economia Italiana." *Industria: Revista di Economia Politica,* no. 4 (1954):463–470.

Roumasset, James. "Rice and Risk: Does Poverty Inhibit Innovation?" Working Paper, no. 44. Davis, California: University of California, 1974.

Rubenstein, Albert H.; Schlie, Theodore W.; and Chakrabarti, Alok K. *Research Priorities on Technology Transfer to Developing Countries: Report of Two Northwestern University/U.S. National Science Foundation Workshops held at Northwestern University, May 1973 and Washington, D.C., September 1973.* Evanston, Ill.: Northwestern University, 1974.

Ruttan, Vernon W. *Technical and Institutional Transfer in Agricultural Development.* New York: Agricultural Development Council, 1973.

Ruttan, Vernon W., and Hayami, Yujiro. *Technology Transfer and Agricultural Development.* New York: Agricultural Development Council, 1973.

Sagasti, Francisco R. "Underdevelopment, Science and Technology: The Point of View of the Underdeveloped Countries: Discussion Paper." *Science Studies* 3 (1973):47–59.

Said, Abdul, and Simons, L. R. *The New Sovereigns.* Englewood Cliffs, N.J.: Prentice Hall, 1975.

Samuelson, P. A. "A Theory of Induced Innovation Along Kennedy-Weizsacker Lines." *Review of Economics and Statistics* LVII (November 1965):343–356.

Schultz, Theodore. *Transforming Traditional Agriculture.* New Haven, Conn.: Yale University Press, 1964.

Schumacher, E. F. *Small is Beautiful: Economics as if People Mattered.* New York: Harper & Row, 1973.

Sen, Amartya K. *Employment, Technology and Development.* Oxford, England: Oxford University Press, 1975.

Service, Elman R. *Primitive Social Organization.* New York: Random House, 1962.

Shaw, G. Keith, and Peacock, Alan. "Possible Means of Promoting Employment by Fiscal Means: Prepared for the Harvard Advisory Group, Indonesia." Cambridge, Mass.: Harvard University, Development Advisory Service, n.d.

Shetty, M. C. *Small-Scale and Household Industries in a Developing Country: A Study of Their Rationale, Structure and Operative Conditions.* New York: Asia Publishing House, 1963.

Smith, Vernon L. *Investment and Production.* Cambridge, Mass.: Harvard University Press, 1961.

Solow, Robert M. "Technical Change and the Aggregate Production Function." *Review of Economics and Statistics* 39 (June 1957):312–320.

Southworth, Herman M., and Johnston, Bruce F., eds. *Agricultural Development and Economic Growth.* Ithaca, New York: Cornell University Press, 1967.

Staley, Eugene, and Morse, Richard. *Small Industry for Developing Countries.* New York: McGraw-Hill, 1965.

Stewart, Frances. "Choice of Techniques in Developing Countries." *The Journal of Development Studies* 9 (October 1972):99–122.

Stiglitz, J. "Alternative Theories of Wage Determination and Unemployment in LDCs: The Labor Turnover Model." Palo Alto, Calif.: Stanford University, 1972.

———. "Alternative Theories of Wage Determination and Unemployment in LDCs: The Efficiency Wage Model." Palo Alto, Calif.: Stanford University, 1972.

———. "Distributions of Wealth and Income Among Individuals." *Econometrica* 37 (July 1969):382–439.

———. *The Efficiency Wage Hypothesis, Surplus Labor, and the Distribution of Income in LDCS.* Stanford University Economics Series, Technical Report, no. 152. Palo Alto, Calif.: Stanford University Press, 1974.

Strassmann, W. Paul, and McConnaughey, John S. "Appropriate Technology for Residential Construction in Less Developed Countries: A Survey of Research Trends and Possibilities." East Lansing, Mich.: Michigan State University, 1972.

Sunkel, Osvaldo. *Past, Present and Future of the Process of Latin-American Development.* Budapest: Hungarian Academy of Sciences, Center for Afro-Asian Research, 1973.

Tax, Sol. *Penny Capitalism: A Guatemalan Indian Economy.* Chicago, Ill.: University of Chicago Press, 1953.

Thorbecke, Eric, ed. *The Role of Agriculture in Economic Development.* New York: National Bureau of Economic Research, 1969.

Timmer, C. Peter; Thomas, John Woodward; Wells, Louis T., Jr.; and Morawetz,

David. *The Choice of Technology in Developing Countries. Some Cautionary Tales.* Cambridge, Mass.: Harvard University, Center for International Affairs, 1975.

Todd, John E. *Size of Firm and Efficiency in Colombian Manufacturing.* Massachusetts Center for Development Economics Research Memorandum, no. 41. Williamstown, Mass.: Williams College, 1971.

Turnham, David. *The Employment Problem in Less Developed Countries.* Paris: Organization for Economic Cooperation and Development, 1971.

United Nations Conference on Trade and Development. *The Role of the Patent System in the Transfer of Technology to Developing Countries: Report prepared jointly by the United Nations Department of Economic and Social Affairs, the UNCTAD Secretariat, and the International Bureau of the World Intellectual Property Organization* (TD/B/AC.11/19/Rev.1.). New York: United Nations Conference on Trade and Development, 1975.

United Nations. Economic Commission for Africa. *African Regional Plan for the Application of Science and Technology to Development* (E/CN.14/579). New York: United Nations, 1973.

United Nations Industrial Development Organization. *Changing Attitudes and Perspectives in Developing Countries Regarding Technology Licensing* (ID/WG.130/3). Vienna: United Nations Industrial Development Organization, 1972.

──────. "Comparable Equipment and Technologies from Developing Countries." Vienna: United Nations Industrial Development Organization, 1975.

──────. "Industrial Equipment from Developing Countries." Vienna: United Nations Industrial Development Organization, 1974.

──────. "Recycling Technologies." Vienna: United Nations Industrial Development Organization, 1974.

──────. *Specification and Remuneration of Foreign Know-How* (ID/WG.130/1). Vienna: United Nations Industrial Development Organization, 1972.

Usui, Mikoto. "Oligopoly, R. and D., and Licensing—A Reflection Towards a Fair Deal in Technology Transfer." Organization for Economic Cooperation and Development, Development Centre, Occasional Paper, no. 7. Paris: Organization for Economic Cooperation and Development, 1975.

Utterback, James M. "The Role of Applied Research Institutes in the Transfer of Technology to Latin America." Bloomington, Indiana: Indiana University, 1973.

Vaitsos, Constantine. "Patents Revisited: Their Function in Developing Countries." *The Journal of Development Studies* 9 (October 1972):71–98.

Vakil, C. N. *Poverty and Planning.* New York: Allied Publishers Private Ltd., 1963.

Wade, Nicholas. "Green Revolution: Creators Still Quite Hopeful on World Food." *Science* 185 (September 6, 1974):844–845.

──────. "New Alchemy Institute: Search for an Alternative Agriculture." *Science* 187 (February 28, 1975):727–729.

Waterston, Albert. *Development Planning: Lessons of Experience.* Baltimore, Md.:, Johns Hopkins Press, 1968.

Weiner, Myron, ed. *Modernization: The Dynamics of Growth.* New York: Basic Books, 1966.

Wells, Louis T., Jr. "Economic Man and Engineering Man: Choice of Technology in Low-Wage Country." In *The Choice of Technology in Developing Countries: Some Cautionary Tales,* by Peter C. Timmer, John Woodward Thomas, Louis T. Wells, Jr., and David Morawetz, pp. 69–86. Cambridge, Mass.: Harvard University, Center for International Affairs, 1975.

White, Lawrence J., ed. *Technology, Employment, and Development: Selected Papers*

Presented at Two Conferences Sponsored by the Council for Asian Manpower Studies. Quezon City, Philippines: Council for Asian Manpower Studies, 1974.

Wilson, J. S. G., and Scheffer, C. T., eds. *Multinational Enterprises.* Leiden, The Netherlands: A. W. Sijthoff, 1974.

Winberg, Nat. "The Multinational Corporation and Labor." In *The New Sovereigns,* by Abdul A. Said and L. R. Simmons. Englewood Cliffs, N.J.: Prentice Hall, 1975.

Wittfogel, Karl A. *Oriental Despotism.* New Haven, Conn.: Yale University Press, 1957.

Contributors

FRANK AHIMAZ, Office of the Dean of Engineering, Cornell University, Ithaca, New York

THOMAS M. ARNDT, Special Assistant for Technology Applications, Bureau for Technical Assistance, U.S. Agency for International Development, Washington, D.C.

HENRY ARNOLD, Director, Office of Science and Technology, Bureau for Technical Assistance, Agency for International Development, Washington, D.C.

JACK BARANSON, President, Developing World Industry and Technology, Washington, D.C.

GERARD K. BOON, El Colegio de Mexico, Mexico City, Mexico

ANTHONY CHURCHILL, Urban Projects Department, International Bank for Reconstruction and Development, Washington, D.C.

JOHN CRAIG, Assistant Director, Denver Research Institute, Denver, Colorado

R. S. DESAI, United Nations Office for Science and Technology, New York, New York

JOHN R. ERIKSSON, Office of Policy Development and Analysis, Bureau for Program and Policy Coordination, Agency for International Development, Washington, D.C.

JOSÉ GIRAL, Department of Chemical Engineering, Faculty of Chemistry, Division of Graduate Studies, National Autonomous University of Mexico, Mexico City, Mexico

CARL D. GODEREZ, Industrial Projects Department, International Bank for Reconstruction and Development, Washington, D.C.

139

FRANZ F. GOLDSCHMIDT, Technical University, Aachen, Germany

ROSS HAMMOND, Industrial Development Division, Engineering Experiment Station, Georgia Institute of Technology, Atlanta, Georgia

HANS HEYMANN, JR., RAND Corporation, Washington, D.C.

SARAH JACKSON, International Economist, Joint Economic Committee, Washington, D.C.

CHARLES L. KUSIK, Arthur D. Little, Inc., Cambridge, Massachusetts

DONALD LEBELL, Office of Academic Planning and Research, University of California, Los Angeles, California

JOHN MCCONNAUGHEY, Building Economics Division, National Bureau of Standards, Washington, D.C.

KEITH MARSDEN, International Labour Organization, Geneva, Switzerland

LEON MILLER, Economic Development Institute, International Bank for Reconstruction and Development, Washington, D.C.

RICHARD MORSE, President, Massachusetts Institute of Technology Development Foundation, Inc., Cambridge, Massachusetts

GUSTAV RANIS, Economic Growth Center, Yale University, New Haven, Connecticut

R. W. RATHJENS, Department of Political Science, Massachusetts Institute of Technology, Cambridge, Massachusetts

JOHN H. REEDY, Arthur D. Little, Inc., Cambridge, Massachusetts

HUGH SCHWARTZ, Industry and Infrastructure Section, Special Studies Division, Inter-American Development Bank, Washington, D.C.

SAADIA M. SHORR, Consultant, Fort Lee, New Jersey

PAUL W. STRASSMANN, Department of Economics, Michigan State University, East Lansing, Michigan

I. DONALD TERNER, Department of Urban Studies and Planning, Massachusetts Institute of Technology, Cambridge, Massachusetts

MAURICIO THOMAE, Education, Science and Technology Section, Project Analysis Department, Inter-American Development Bank, Washington, D.C.

MICHAEL P. TODARO, Technology and Employment Project, Rockefeller Foundation, New York, New York

JOSÉ VILLAVICENCIO, Arthur Young and Co., Washington, D.C.

STEPHEN WEED, Western Africa Regional Projects Department, International Bank for Reconstruction and Development, Washington, D.C.

CHARLES WEISS, JR., Science and Technology Adviser, International Bank for Reconstruction and Development, Washington, D.C.